Collins Revision

GCSE Foundation Physics

Revision Guide

FOR OCR GATEWAY B

About this book

This book covers GCSE Physics for OCR Gateway B at Foundation Level. Written by GCSE examiners, it is designed to help you to get the best grade in your GCSE Physics exams.
The book is divided into three parts; a topic-by-topic revision guide, workbook practice pages for each topic and detachable answers.

How to Use It
The revision guide section gives you complete coverage of each of the six modules that you need to study. Use it to build your knowledge and understanding.
The workbook section is packed with exam-style questions. Once you have covered a particular topic, use the matching workbook page to test yourself.
The answers in the back of the book are detachable. Remove them to help check your grade or a friends.

Go Up a Grade
There are lots of revision guides for you to choose from. This one is different because it really helps you to go up a grade. Each topic in the revision guide and workbook sections is broken down and graded to show you what examiners look for at each level. This lets you check where you are, and see exactly what you need to do to improve your grade at every step. Crucially, it shows you what makes the difference between an E–F and a C–D grade answer.

Special Features
- **Questions** at the end of every topic page quickly test your level.
- **Top Tips** give you extra advice about what examiners really want.
- **Summaries** of each module remind you of the most important things to remember.
- **Checklists** for each module help you to monitor your progress.
- A comprehensive **glossary** gives you a quick reference guide to the Physics terms that you need to know.

Published by Collins
An imprint of HarperCollins*Publishers*
77–85 Fulham Palace Road
Hammersmith
London W6 8JB

Browse the complete Collins catalogue at
www.collinseducation.co.uk

© HarperCollins*Publishers* Limited 2010

10 9 8 7 6 5 4 3 2 1

ISBN-13 978-0-00-734810-7

The authors assert their moral rights to be identified as the authors of this work.

All rights reserved. No part of this publication may be reproduced, stored in a retrieval system, or transmitted in any form or by any means, electronic, mechanical, photocopying, recording or otherwise, without the prior written permission of the Publisher or a licence permitting restricted copying in the United Kingdom issued by the Copyright Licensing Agency Ltd., 90 Tottenham Court Road, London W1T 4LP.

British Library Cataloguing in Publication Data
A Catalogue record for this publication is available from the British Library

Written by Sandra Mitchell and Chris Sherry
Series Consultant Chris Sherry
Project Manager Charis Evans
Design and layout Graham Brasnett
Editor Mitch Fitton
Illustrated by Kathy Baxendale, IFA design Ltd, Mark Walker, Bob Lea and Steve Evans
Indexed by Marie Lorimer
Printed and bound in the UK by Martins the Printers, Berwick Upon Tweed

Acknowledgements
The Authors and Publishers are grateful to the following for permission to reproduce photographs:

Biophoto Associates p7/Corbis p9/Jeremy Walker p13/ Photos.com p20, p25, p29T, p88/David Parker p26/ Lynton & Lynmouth Cliff Railway p29T/G. Brad Lewis p36/ P. Fowler & D. Perkins p37/American Institute of Physics p42T/Hermann Eisenbeiss p42B, Jupiter Images p43C/ Glynis Kirk, Action Plus p43B/Carol & Mark Werner, Phototake p45, istockphoto p56, p58, p76/Steve Gschmeissner p91/Edward Kinsman p92.

Whilst every effort has been made to trace the copyright holders, in cases where this has been unsuccessful, or if any have inadvertently been overlooked, the Publishers will be pleased to make the necessary arrangements at the first opportunity.

Contents

P1 Energy for the home

	Revision	Workbook
Heating houses	4	62
Keeping homes warm	5	63
How insulation works	6	64
Cooking with waves	7	65
Infrared signals	8	66
Wireless signals	9	67
Light	10	68
Stable Earth	11	69
P1 Summary	12	70

P2 Living for the future

	Revision	Workbook
Collecting energy from the Sun	13	71
Generating electricity	14	72
Fuels for power	15	73
Nuclear radiations	16	74
Our magnetic field	17	75
Exploring our Solar System	18	76
Threats to Earth	19	77
The Big Bang	20	78
P2 Summary	21	79

P3 Forces for transport

	Revision	Workbook
Speed	22	80
Changing speed	23	81
Forces and motion	24	82
Work and power	25	83
Energy on the move	26	84
Crumple zones	27	85
Falling safely	28	86
The energy of games and theme rides	29	87
P3 Summary	30	88

P4 Radiation for life

	Revision	Workbook
Sparks!	31	89
Uses of electrostatics	32	90
Safe electricals	33	91
Ultrasound	34	92
Treatment	35	93
What is radioactivity?	36	94
Uses of radioisotopes	37	95
Fission	38	96
P4 Summary	39	97

P5 Space for reflection

	Revision	Workbook
Satellites, gravity and circular motion	40	98
Vectors and equations of motion	41	99
Projectile motion	42	100
Momentum	43	101
Satellite communication	44	102
Nature of waves	45	103
Refraction of waves	46	104
Optics	47	105
P5 Summary	48	106

P6 Electricity for gadgets

	Revision	Workbook
Resisting	49	107
Sharing	50	108
Motoring	51	109
Generating	52	110
Transforming	53	111
Charging	54	112
It's logical	55	113
Even more logical	56	114
P6 Summary	57	115

How Science Works	58–9
Glossary	116–20
Answers	121–8

Heating houses

Heat

- **Hot** objects have a high temperature and usually cool down; cold objects have a lower temperature and usually warm up.
- The unit of **temperature** is the **degree Celsius** (°C).
- **Heat** is a form of energy. The unit of energy (heat) is the **joule** (J).
- Energy, in the form of heat, flows from a warmer to a colder body. When energy flows away from a warm object, the temperature of that object decreases.
- Temperature is a measure of 'hotness'. It allows one object to be compared to another.

Specific heat capacity

- The **energy** needed to change the temperature of a substance depends on:
  - **mass**
  - the **material** it's made from
  - the **temperature change**.
- When you heat a solid, its temperature rises until it changes to a liquid. The temperature stays the same until all of the solid has changed to a liquid. The temperature of the liquid then rises until it changes into a gas. The temperature stays the same until all of the liquid has changed to a gas.
- Heat is needed to **melt** a solid at its **melting point**.
- Heat is needed to **boil** a liquid at its **boiling point**.
- All substances have a property called **specific heat capacity**, which is:
  - the energy needed to raise the temperature of 1 kg by 1 °C
  - measured in joule per kilogram degree Celsius (J/kg°C)
  - different for different materials.
- When an object is heated and its temperature rises, energy is transferred.

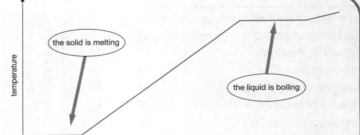

Why does the line flatten out during changes of state?

Specific latent heat

- **Latent heat** is the heat needed to change the state of a material without a change in temperature.
- **Specific latent heat** is:
  - the energy needed to melt or boil 1 kg of the material
  - measured in joule per kilogram (J/kg)
  - different for different materials and each of the changes of state.
- When an object is heated and it changes state, energy is transferred.

Top Tip!
Remember that heat and temperature are different. Energy transfer doesn't always involve a rise or fall in temperature.

Questions

Grades G-E
1 What is the difference between temperature and heat?

Grades D-C
2 When you put an ice cube on your hand, your hand gets colder. Suggest why.

Grades G-E
3 Describe what happens to a solid at its melting point.

Grades D-C
4 The syrup in a steamed syrup pudding always appears to be hotter than the sponge even though they're at the same temperature. Why is this?

Keeping homes warm

Energy

- **Insulation** reduces energy loss from a home.
- Insulation contains a lot of **trapped air** because air is a good insulator.
- Energy loss from uninsulated homes can be reduced by:
  - injecting **foam** into the cavity between the inner and outer walls
  - laying **fibreglass** or a similar material between the joists in the loft
  - replacing single glazed windows with **double** (or triple) **glazed windows**
  - drawing the **curtains**
  - placing **shiny foil** behind radiators
  - **sealing gaps** around doors.

Top Tip! Remember, it's the **trapped air** that provides insulation.

- Different types of insulation cost different amounts and save different amounts of energy.
- Most energy is lost from the **walls** of an uninsulated house.
- To work out the most cost-effective type of insulation, the **payback time** is calculated:

$$\text{payback time} = \frac{\text{cost of insulation}}{\text{annual saving}}$$

Draught proofing costs £80 to install and saves £20 per year on heating bills.

$$\text{payback time} = \frac{80}{20}$$
$$= 4 \text{ years}$$

Would energy be saved if the aluminium foil was fixed to the front of radiators?

- Different energy sources each have their advantages and disadvantages. Cost can be compared by using a consistent unit – **kWh**.
- The formula for energy efficiency is:

$$\text{efficiency} = \frac{\text{useful energy output}}{\text{total energy input}}$$

For every 100 J of energy in coal, 27 J are transferred to a room as heat by a coal fire.

$$\text{efficiency} = \frac{27}{100}$$
$$= 0.27 \text{ or } 27\%$$

- Coal fires are very inefficient because so much heat is *lost* via the chimney.

Top Tip! Most of our energy sources come from burning fossil fuels. This contributes to climate change as the average home emits 7 tonnes of carbon dioxide every year.

Questions

Grades G-E
1. Why is fibreglass a good material to insulate a loft?
2. Why does shiny foil behind a radiator help to reduce energy loss?

Grades D-C
3. Doug spends £3000 fitting double glazing. It saves him £50 each year on his heating bills. Calculate the payback time.
4. A coal fire has an efficiency of 0.35 (35%). Mrs Tarantino spends £150 per year on coal. How much money does she waste heating the surroundings instead of the room?

How insulation works

Insulators

- Air is a good **insulator** because it doesn't allow energy to transfer from a warm body to cooler surroundings.
- Fur coats keep us warm because they have a lot of trapped air.

Convection and radiation

- **Convection** in air takes place when hot air rises and cooler air falls to take its place.
- This movement of air is called a **convection current**.
- **Infrared radiation** from the Sun can be reflected by a shiny surface. The heat can be used for cooking or producing electricity.

House insulation

- Double glazing reduces energy loss by **conduction**. The gap between the two pieces of glass is filled with a gas or contains a **vacuum**.
- **Solids** are good conductors because particles are close together. They can transfer energy easily.

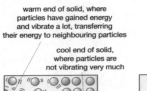
warm end of solid, where particles have gained energy and vibrate a lot, transferring their energy to neighbouring particles
cool end of solid, where particles are not vibrating very much
a solid is a good conductor

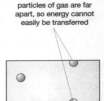
particles of gas are far apart, so energy cannot easily be transferred
air and argon are good insulators

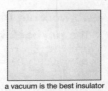
no particles, so no energy transfer
a vacuum is the best insulator

Why is air a better insulator than a solid?

- Particles in a **gas** are far apart, so it's difficult to transfer energy.
- There are no particles in a **vacuum**, so it's impossible to transfer energy by conduction.
- **Loft insulation** reduces energy loss by **conduction** and **convection**:
  - warm air in the home rises
  - energy is transferred through the ceiling by conduction
  - air in the loft is warmed by the top of the ceiling
  - the warm air is trapped in the loft insulation
  - both sides of the ceiling are at the same temperature so no energy is transferred.

snow does not melt on a well insulated house
cold air in loft
loft insulation
ceiling
warm air in room

- Without loft insulation:
  - the warm air in the loft can move by convection and heat the roof tiles
  - energy is transferred to the outside by conduction.
- **Cavity wall insulation** reduces energy loss by conduction and convection:
  - the air in the foam is a good insulator
  - the air can't move by convection because it's trapped in the foam.
- **Insulation blocks** used to build new homes have shiny foil on both sides so:
  - energy from the Sun is reflected back to keep the home cool in summer
  - energy from the home is reflected back to keep the home warm in winter.

Top Tip! Remember, hot air will only rise into the loft if the loft-hatch is open.

Questions

(Grades G-E)
1 Why does a woollen jumper keep us warm?
2 Write down two ways in which radiated energy from the Sun can be used.

(Grades D-C)
3 Suggest why a vacuum-sealed double glazed unit is better than having thicker glass in a window.
4 Debbie says that snow doesn't melt as quickly on the roof of a well-insulated house. Explain why.

Cooking with waves

Infrared and microwave radiation

- The **electromagnetic spectrum** is a family of waves that have different **frequencies** and **wavelengths**. **Microwaves** and **infrared** radiation are part of the **electromagnetic spectrum**. They have wavelengths that are **longer** than the wavelength of visible light.

- All warm and hot bodies **emit infrared** radiation. Hotter bodies emit more radiation than cooler bodies.
  – Dark, dull surfaces emit more radiation than light shiny surfaces.

- Infrared radiation is **absorbed** by all bodies, which then become warmer.
  – Dark, dull surfaces absorb more radiation than light shiny surfaces.

- Water molecules in food absorb microwaves.

- Microwaves penetrate up to 1 cm into food but infrared radiation doesn't penetrate food very easily.

- Microwaves can penetrate glass or plastic but are reflected by shiny metal surfaces. The door of a microwave oven is made from a special reflective glass.

Top Tip! Black surfaces absorb radiation; they don't attract heat.

Communication

- **Mobile phones** use microwaves for communication.

- There's some evidence that microwaves from mobile phones can affect the body. Young people are advised not to use mobile phones unless they have to.

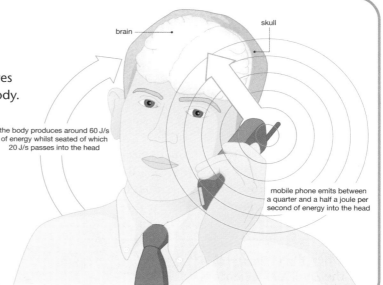

the body produces around 60 J/s of energy whilst seated of which 20 J/s passes into the head

mobile phone emits between a quarter and a half a joule per second of energy into the head

The science of mobile phones.

- Microwave radiation is used to communicate over long distances. The transmitter and receiver must be in **line of sight**. Aerials are normally situated on the top of high buildings.

- **Satellites** are used for microwave communication. The signal from Earth is received, amplified and re-transmitted back to Earth. Satellites are in line of sight because there are no obstructions in space. The large aerials can handle thousands of phone calls and television channels at once.

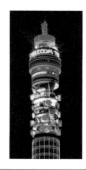

The Telecom Tower. Why are microwave aerials on high buildings?

Questions

Grades G-E
1. Which emits most infrared radiation – an electric fire or an electric iron?

Grades D-C
2. Suggest why microwave ovens need to have a special glass door.

Grades G-E
3. How can young people use a mobile phone more safely?

Grades D-C
4. Some areas of the country have very poor mobile phone reception. Suggest why.

Infrared signals

Signals

- **Infrared signals** can be used for **remote controllers**, such as television and other audio-visual equipment; a cordless mouse; and household equipment like garage doors and blinds.
- **Passive infrared sensors** detect body heat and are used in burglar alarms.
- Signals can be **digital** or **analogue**.
- **Digital** signals have two values – **on** and **off**.
- **Analogue** signals can have any value and are continuously **variable**. The analogue signal changes both its **amplitude** and **wavelength**.

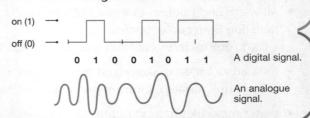

A digital signal.

An analogue signal.

Total internal reflection and critical angle

- Light can be reflected inside materials such as glass, water and perspex. This process is called **total internal reflection**.
- Laser beams and infrared radiation can travel along optical fibres.
- Telephone conversations and computer data are transmitted long distances along **optical fibres**.
- Some fibres are coated to improve reflection.
- When light travels from one material to another it's **refracted**.
- If light is passing from a more dense material into a less dense material, the **angle of refraction** is larger than the **angle of incidence**.
- When the angle of refraction is 90°, the angle of incidence is called the **critical angle**.
- If the angle of incidence is **bigger** than the critical angle, the light is reflected. This is **total internal reflection**.

Top Tip! Remember optical fibres are solid, not hollow!

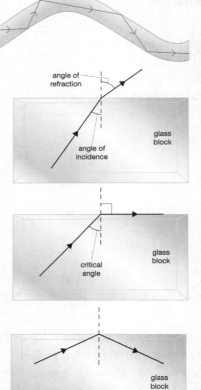

What happens to light when it meets the boundary with air in this optical fibre?

Questions

(Grades G-E)
1. What type of radiation is produced when you use the remote control for your DVD player?

(Grades D-C)
2. Draw a diagram of the scale of an analogue voltmeter that can read up to 5 V.

(Grades G-E)
3. Write down two types of radiation that can travel along an optical fibre.

(Grades D-C)
4. The critical angle for light passing from glass into air is 41°. Draw a diagram to show what happens if the angle of incidence is 30°.

Wireless signals

Radio waves

- **Wireless technology** means that computers and phones can be used anywhere because they're portable, which makes them very convenient.

- Wireless technology means what it says – there are no wires; radio waves are used instead.
- **Radio waves** are part of the **electromagnetic spectrum**.
- Radio waves can be reflected, which means that a signal can be received even if the aerial is not in line of sight.
- Sometimes an **aerial** receives a reflected signal as well as a direct signal. On a television picture this appears as **ghosting**.

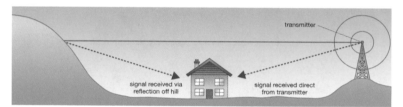

Radio waves can be reflected off solid obstacles such as a hill.

- Radio waves are refracted in the upper atmosphere. The amount of **refraction** depends on the frequency of the wave. There's less refraction at higher frequencies.

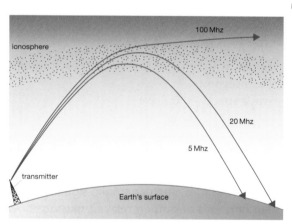

What type of wave shows most refraction?

- Radio stations broadcast signals with a particular frequency. The same frequency can be used by more than one radio station because the distance between the stations means that only one will be received. However, in unusual weather conditions, the radio waves can travel further and the broadcasts **interfere**.

Questions

Grades G-E

1. What's the main advantage of wireless technology?

Grades D-C

2. Radio stations that broadcast on FM use frequencies of about 100 MHz or higher. Suggest why there's less interference when you listen to a radio station broadcasting on FM.

Light

Transverse waves

- When a pebble is dropped into a pool of water, a **circular wave** spreads out. The water particles move up and down as the wave spreads out.
- Water waves are **transverse** waves. A transverse wave travels at **right angles** to the wave vibration.
- The speed of light, and all waves in the electromagnetic spectrum, is 300 000 km/s.

Wave properties

- The **amplitude** of a wave is the maximum displacement of a particle **from** its rest position.
- The **trough** of a wave is the displacement of a particle **below** its rest position.
- The **crest** of a wave is the displacement of a particle **above** its rest position.
- The **wavelength** of a wave is the distance **between** two adjacent points of similar displacement on the wave.
- The **frequency** of a wave is the **number of complete waves** passing a point in one second.

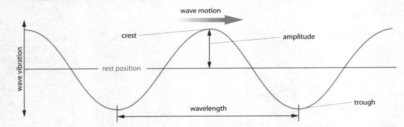

The parts of a transverse wave.

- The formula for calculating wave speed is:
  wave speed = frequency × wavelength

  When Katie throws a stone into a pond, the distance between ripples is 0.3 m and four waves reach the edge of the pond each second.
  wave speed = 0.3 × 4
             = 1.2 m/s

Top Tip!
Remember to always give the units in your answer:
– wavelength – **metre** (m)
– frequency – **hertz** (Hz)
– speed – **metre per second** (m/s)

Sending messages

- Early messages, such as smoke signals, beacons and semaphore, relied on **line of sight**. Runners and horsemen relayed messages over longer distances.
- Signalling lamps need a code to represent letters.
- Lasers produce a very intense beam of light. Laser light can be used along optical fibres for communication.
- The **Morse code** uses a series of dots and dashes to represent letters of the alphabet. This code is used by signalling lamps as a series of short and long **flashes of light**.

Questions

Grades G-E
1. What's the speed of infrared radiation?

Grades D-C
2. A sound wave has a frequency of 200 Hz and a wavelength of 1.5 m. Calculate the speed of sound.

Grades G-E
3. What's the advantage of using a signalling lamp instead of sending a runner?

Grades D-C
4. How does Morse code work?

Stable Earth

Earthquakes

- Earthquakes happen at a **fault**.
- **Shock waves** travel through and round the surface of the Earth.
- We detect earthquakes using a **seismometer**.
  - A heavy weight, with a pen attached, is suspended above a rotating drum with graph paper on it. The base is bolted to solid rock.
  - During an earthquake, the base moves and the pen draws a trace on the graph paper.
- The **focus** is where the earthquake happens below the surface.
- The **epicentre** is the point on the surface above the focus.
- **L waves** travel round the surface very slowly.
- **P waves** are **longitudinal** pressure waves:
  - P waves travel through the Earth at between 5 km/s and 8 km/s
  - P waves can pass through solids and liquids.
- **S waves** are **transverse** waves:
  - S waves travel through the Earth at between 3 km/s and 5.5 km/s
  - S waves can only pass through solids.
- A **seismograph** shows the different types of earthquake wave.

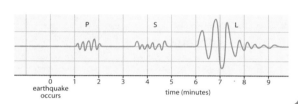

A seismograph. Which type of wave travelled fastest during the earthquake?

Weather effects

- The Earth is getting warmer because there are more **greenhouse gases** in the atmosphere.
- **Carbon dioxide** is the main greenhouse gas produced whenever a fuel burns. The more energy we use, the more carbon dioxide there'll be.
- Trees and plants use carbon dioxide *but* large areas of **forest** are being destroyed.
- **Natural events** and **human activity** affect our weather.
  - Dust from **volcanoes** can reflect energy from the Sun back into the atmosphere – it becomes cooler on Earth.
  - Dust from **factories** can reflect radiation from towns back to Earth – it becomes warmer on Earth.
- **Ultraviolet** radiation causes suntan. Too much exposure to ultraviolet radiation can cause sunburn or skin cancer. Sunscreen filters out ultraviolet radiation.
- Ultraviolet light on the skin causes the cells to make **melanin**, a pigment that produces a tan. People with dark skin don't tan easily because ultraviolet radiation is filtered out.
- Use a sunscreen with a high **SPF** (sun protection factor) to reduce risks. The formula for calculating a safe length of time to spend in the Sun is:
  safe length of time to spend in the Sun = published normal burn time × SPF

Questions

Grades G-E
1 What's a seismometer?

Grades D-C
2 What's the difference between the focus and the epicentre of an earthquake?

Grades G-E
3 Write down two ways in which human activity is contributing to climate change.

Grades D-C
4 Kate sees that on a hot day the normal burn time is 5 minutes. How long can she stay in the sun, without burning, if she uses sunscreen with SPF 30?

P1 Summary

P1 ENERGY FOR THE HOME

Heat and temperature

Hot objects have high temperatures and tend to cool down.
Cold objects have low temperatures and tend to warm up.
Energy is **transferred** from a hotter to a colder object.

Temperature is a measure of **hotness** in degree Celcius (°C). Heat is a measure of **energy transfer** in joules (J).

Energy is **transferred** when a substance **changes temperature**. The amount of energy transferred depends on:
– the mass
– temperature change
– specific heat capacity.

Energy is **transferred** when a substance **changes state**. The amount of energy transferred depends on:
– the mass
– the specific latent heat.

Energy transfer

Air is a good **insulator** and reduces energy transfer by **conduction**.

Trapped air reduces energy transfer by **convection**.

Shiny surfaces **reflect** infrared radiation to reduce energy transfer.

$$\text{efficiency} = \frac{\text{useful energy output}}{\text{total energy input}}$$

Energy saving in the home can be achieved by:
– double glazing
– cavity wall insulation
– draught strip
– reflecting foil
– loft insulation
– curtains
– careful design.

Waves carrying energy

Warm and hot objects emit **infrared radiation**. Infrared radiation is used for cooking and remote controllers.

Microwaves can be used for **cooking** and for **communication** when transmitter and receiver are in line of sight.

Digital and **analogue** signals are used for **communication**.

Radio waves are used for **communication**.

Waves in the electromagnetic spectrum are:
– **radio waves**
– **microwaves**
– **infrared**
– **visible light**
– **ultraviolet**.
All electromagnetic waves can be **reflected** and **refracted**.

The stable Earth

Earthquakes occur at plate boundaries. Earthquake waves travel through the Earth.

Exposure to **ultraviolet radiation** causes **sun burn** and **skin cancer**. Sunscreen and sunblock reduce damage caused by ultraviolet waves.

Climate change is a result of both **human activity**, such as burning fossil fuels, and **natural phenomena**, such as volcanoes.

Collecting energy from the Sun

Unlimited energy

- The **Sun** is the **renewable** source of energy for the Earth.
- **Light** from the Sun allows plants to **photosynthesise** and **heat** from the Sun provides the warmth living things need.

Photocells

- **Photocells** use **light**, and **solar cells** use **light from the Sun** to produce **direct current** (DC) electricity. This means they can be used in remote locations.
  - The **larger** the area of the **photocell**, the more electricity is produced.
  - The **greater** the **light intensity**, the more electricity is produced.

Why do you think there's a photocell on top of the parking meter?

- The **advantages** of **photocells** are:
  - they're robust and don't need much maintenance
  - they don't need any fuel or long power cables
  - they don't cause pollution or contribute to climate change
  - they use a renewable energy resource.
- The only **disadvantage** is that they don't produce electricity when it's dark.

Renewable energy

- **Solar panels** use energy from the Sun to heat water. Solar panels are black to absorb as much **radiation** as possible and the warmed water in their tubes passes to a storage tank by **convection**.
- **Convection currents** are formed in the air when there are differences in land temperature. This is known as **wind**.
- Wind can be used to turn a **wind turbine** and produce **electricity**.

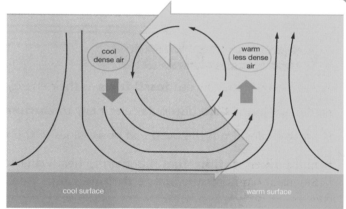

What causes a convection current?

- Moving air has **kinetic energy** which is transferred into electricity by a wind turbine.
- A house that uses **passive solar heating** makes use of direct sunlight. It has large windows facing the Sun (South) and small windows facing North.
  - During the **day**, energy from the Sun warms the walls and floors.
  - During the **night**, the walls and floors radiate energy back into the room.
- Curved solar reflectors **focus** energy from the Sun.

Top Tip!
In the Southern Hemisphere, the larger windows will need to face North.

Questions

Grades G-E
1. What's the difference between a photocell and a solar cell?

Grades D-C
2. Some roadside warning signs are powered using photocells. Suggest why.

Grades G-E
3. Why is a solar panel coloured black?

Grades D-C
4. Why do the large windows in a solar-heated house have to face North in Australia but South in England?

Generating electricity

The dynamo effect

- A **dynamo** is one example of a **generator**. A **magnet** rotates inside a coil of wire to produce **alternating current** (AC).
- If a wire is moved near a magnet, or vice versa, alternating current is produced in the wire.
- The **current** from a dynamo can be increased by:
  - using a **stronger** magnet
  - **increasing** the number of **turns** on the coil
  - **rotating** the magnet **faster**.
- The voltage and frequency from a dynamo can be displayed on an **oscilloscope**.
- The formula to work out the frequency is:
  frequency (in hertz, Hz) = 1 ÷ period (in seconds, s)

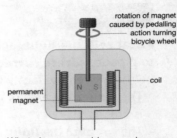

What changes could you make to this dynamo to increase the current output?

Generators

- The **generator** at a power station works like a dynamo.
- A simple generator consists of a coil of wire rotating between the poles of a magnet to produce a current in the coil.

Top Tip! It's the relative movement of the magnet and coil that's important.

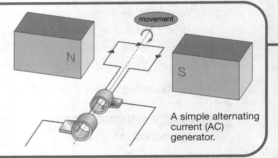

A simple alternating current (AC) generator.

Distributing electricity

- Most power stations use **fossil fuels** as their energy source.
- The **National Grid** distributes electricity to **consumers**.
- Energy is lost from the overhead power lines to the surroundings as heat.
- In a **power station**, **fuel** is burned to heat water to produce **steam**. Steam at high pressure turns a **turbine** which then drives a **generator**.

Transformers

- A **transformer** changes the size of an alternating voltage.
- Some transformers increase (**step-up**) voltages, others decrease (**step-down**) voltages.

Top Tip! A transformer *doesn't* change alternating current to direct current.

- The **National Grid** distributes electricity around the country at high voltages. This means that:
  - there's less energy loss
  - the distribution costs are lower
  - electricity prices are cheaper.

Questions

Grades G-E

1. What type of current is produced by **a** a battery **b** a dynamo?
2. Write down two examples of fossil fuels used in power stations.

Grades D-C

3. Suggest why steam is used to turn a turbine at most power stations.
4. What are the advantages of distributing electricity around the country at high voltages?

Fuels for power

Energy sources

- Most **power stations** use **non-renewable fossil fuels** as their energy source: coal, gas and oil.
- Some modern power stations use **renewable biomass** as their energy source: wood, straw and manure.
- **Nuclear power stations** use **non-renewable uranium** as their energy source.
- A **fuel** burns in air to release energy in the form of heat.
- **Biomass** can be burned. It's usually allowed to **ferment** to produce **methane**, which is then burned.
- **Nuclear fuels** don't burn. In a nuclear reactor, uranium atoms split and release lots of energy in the form of heat.
- Nuclear reactions in power stations have to be controlled to avoid an explosion.

Nuclear waste

- **Nuclear power stations** don't produce carbon dioxide or smoke, and they don't contribute to climate change.
- Waste from a nuclear power station is **radioactive**.
  – Low-level radioactive waste can be pumped out to sea.
  – High-level waste has to be stored in steel drums deep underground.
- **Radiation** from nuclear waste causes **ionisation**, which causes a change in the structure of any atom exposed to the radiation.
  – The cells of our bodies are made of many different atoms. DNA is an important chemical in a cell and can be changed when it's exposed to radiation. The cell behaves differently to normal and this is called **mutation**.
  – One effect of mutation is for a cell to divide in an uncontrolled way. This can lead to **cancer**.
- **Waste** from a nuclear reactor can remain radioactive for thousands of years.
- **Plutonium**, one of the waste products, can be used to make nuclear weapons. Nuclear bombs destroy everything in a very large area and make that area unusable for a long time.

Power

- **Power** is a measure of the rate at which energy is used. The unit is the **watt** (W). The formula for power is: power = voltage × current
- **Electrical consumption** is the amount of energy that's been used. The unit is the **kilowatt hour** (kWh). The formula for energy used is:
  energy used = power × time
- The formula to work out the **cost** of using an electrical appliance is:
  cost of electricity used = energy used × cost per kWh

Questions

Grades G-E
1. What's meant by a non-renewable energy source?
2. Write down one advantage of producing electricity from a nuclear power station.

Grades D-C
3. Explain how exposure to nuclear radiation can cause cancer.
4. How much does it cost to use a 2 kW toaster for $\frac{1}{4}$ hour if electricity costs 10p per kWh?

Nuclear radiations

Background radiation

- The nuclear radiation that's always around us is called **background radiation**.
- Radioactivity is measured with a Geiger-Müller (GM) tube connected to a **ratemeter** – a **Geiger counter**.
- Most background radiation is naturally occurring.

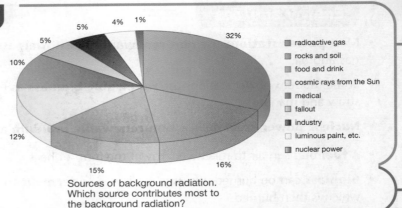

Sources of background radiation. Which source contributes most to the background radiation?

Properties of ionising radiations

- The three main types of **ionising radiations** are: **alpha** (α), **beta** (β) and **gamma** (γ).
- Alpha, beta and gamma radiations come from the nucleus of an atom. **Alpha** radiation causes the **most** ionisation and **gamma** radiation the **least**. Ionisation produces **charged particles**.

type of ionising radiation	range	penetration through materials
alpha	short – a few centimetres	easily absorbed by a sheet of paper or card
beta	about 1 metre	absorbed by a few millimetres of aluminium
gamma	theoretically infinite	can pass through metres of lead or concrete

Handling radioactivity

- When handling radioactivity, you should:
  – wear suitable protective clothing
  – not handle it directly
  – keep a safe distance: handle with tongs
  – use shielding to absorb radiation
  – use for the minimum time necessary.
- **Radioactive waste** is disposed of at sea, buried in landfill sites or stored deep underground.

Uses of radioactivity

- Ionising radiation kills cells and living organisms. **Cobalt-60** is a radioactive material used to treat **cancers**.
- Radiation kills bacteria and microbes, so medical instruments are **sterilised** by gamma radiation.
- Some radioactive waste can be **reprocessed** into new, useful radioactive material.
- **Alpha radiation** is used in **smoke alarms**.
- **Beta** or **gamma** sources are used in **rolling mills** to conrol **thickness**.
- **Gamma** sources are injected into the body as **tracers**.

Questions

Grades G-E
1. What is meant by the term 'background radiation'?

Grades D-C
2. What fraction of background radiation comes from natural sources?

Grades G-E
3. Suggest two reasons why a school teacher always handles a radioactive source with tongs.

Grades D-C
4. A rolling mill uses beta radiation to control the thickness of cardboard. Explain why alpha radiation and gamma radiation aren't used.

Our magnetic field

Magnetic fields

- The Earth behaves as if it has a large magnet at its centre. The **magnetic field** around the Earth is similar to the field due to a **bar magnet**.
- The Earth's magnetic field is strongest at the North and South **poles**.
- A **compass** points towards magnetic North along the direction of the magnetic field.
- The **outer core** of the Earth is mainly **molten iron**; it's too hot to be magnetic.
- An **electric current** in a wire or a coil has a magnetic field around it too.
- The magnetic field around a coil of wire is similar to the field around a bar magnet.
- A magnet shouldn't be brought close to a **television** or **computer monitor** because it causes the electron beam in the tube to change direction. The beam strikes the wrong part of the screen giving a distorted picture.

Top Tip! The model of Earth's magnetic field has a magnetic South pole at geographic North.

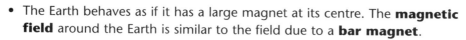

Compare the shape of the magnetic fields around
a a bar magnet and
b the Earth.

Satellites

- The Moon is a **natural satellite** that orbits the Earth.
- The Moon and the Earth were probably formed by two planets colliding 4.6 billion years ago.
- **Artificial satellites** orbit the Earth. They provide information about **weather**, are used as **navigation aids**, and for **communication** and **spying**.

Solar effects

- The **Sun** is a source of **ionising radiation**.
- **Solar flares** erupt from the Sun's surface and cause disruption to communication signals.
- The **energy** from a solar flare is equivalent to that from a million hydrogen bombs.
- Large numbers of charged particles are emitted at very high speeds. These produce magnetic fields that interact with the Earth's magnetic field.

Origin of the Moon

- When the Solar System was formed, there were probably more planets than today.
- Scientists believe there was another planet in the same orbit as the Earth.
  - The two planets collided and were both almost totally destroyed.
  - Iron became concentrated at the core of the new Earth, less dense rocks started to orbit.
  - These rocks clumped together to form the Moon.
  - The Earth and the Moon were rotating much faster than they do now – the Moon caused the speed of rotation to slow down.

Questions

Grades G-E
1. Draw a diagram to show the magnetic field around a bar magnet. Show clearly the magnetic poles and the direction of the magnetic field.

Grades D-C
2. A television screen is coated with red, blue and green dots. What happens to the dots when they're hit by a beam of electrons?

Grades G-E
3. What's a satellite?

Grades D-C
4. How do scientists believe the Moon was formed?

Exploring our Solar System

The Universe

- Our **galaxy** is called the **Milky Way**, and the Milky Way is one of the galaxies that make up the **Universe**.
- **Stars** are very hot and produce their own light – that's why we can see them.
- Sometimes we can see **meteors**, **comets** and **orbiting satellites** because they reflect light.
- **Meteorites** are large rocks that don't burn up as they fall to Earth.
- We can't see **black holes**.
- The Earth is one of the planets in our Solar System orbiting the Sun.
- There may be another planet, three times further away than Pluto and larger.
- In August 2006, scientists decided that Pluto shouldn't be considered a planet because of its size and orbit shape.

Top Tip!
Use a mnemonic to remember the order of the planets: **M**any **V**ets **E**arn **M**oney **J**ust **S**itting **U**nder **N**uts (**P**lants).

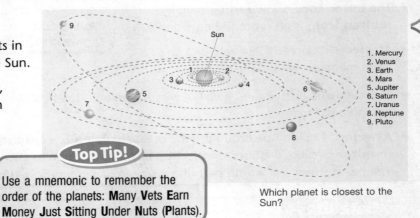

1. Mercury
2. Venus
3. Earth
4. Mars
5. Jupiter
6. Saturn
7. Uranus
8. Neptune
9. Pluto

Which planet is closest to the Sun?

Exploration

- **Radio signals** have been sent into space since 1901.
- A coded signal about life on the Earth was sent towards another star system in 1974 – it will be at least 40 000 years before we can expect a reply!
- Spacecraft showing pictures of life on the Earth and containing records of sounds from the Earth have also been sent into space.
- Unmanned spacecraft (**probes**) have explored the surface of the Moon and Mars. They can go to places where humans can't survive.
- The **Hubble Space Telescope** orbits the Earth collecting information from distant galaxies.
- The Moon is the only body in space visited by humans. **Astronauts** can wear normal clothing in a pressurised spacecraft, but outside the spacecraft they need to wear special **spacesuits**.
  – A dark visor stops an astronaut being blinded.
  – The suit is pressurised and has an oxygen supply for breathing.
  – The surface of the suit facing towards the Sun can reach 120 °C.
  – The surface of the suit facing away from the Sun may be as cold as −160 °C.
- When travelling in space, astronauts experience **lower gravitational forces** than on the Earth.

It will be a long time before we can expect a reply!

Questions

Grades G-E
1 Write down three things we can see in the sky because they reflect light.

Grades D-C
2 Write down the names of the planets in the Solar System in order from the Sun.

Grades G-E
3 How many years will it take for the radio message, sent in 1974, to reach the star system at which it was aimed?

Grades D-C
4 Why do astronauts have to wear special spacesuits when outside their spacecraft?

Threats to Earth

Asteroids

- When an **asteroid** hits the Earth, it leaves a large **crater**.
- The asteroid that collided with the Earth 65 million years ago caused more than just a crater:
  - hot rocks rained down
  - there were widespread fires
  - **tsunamis** flooded large areas
  - clouds of dust and water vapour spread around the world in the upper atmosphere
  - sunlight couldn't penetrate and temperatures fell
  - 70 per cent of all species, including dinosaurs, became extinct.
- Asteroids are **mini-planets** or **planetoids** orbiting the Sun in a 'belt' between Mars and Jupiter. They're large rocks that were left over from the formation of the Solar System.
- **Geologists** have examined evidence to support the theory that asteroids have collided with the Earth.
  - Near to a crater thought to have resulted from an asteroid impact, they found quantities of the metal **iridium** – a metal not normally found in the Earth's crust but common in meteorites.
  - Many **fossils** are found below the layer of iridium, but few fossils are found above it.
  - **Tsunamis** have disturbed the fossil layers, carrying some fossil fragments up to 300 km inland.

Comets

- **Comets** are lumps of ice filled with dust and rock.
- When a comet orbits close to the Sun, the ice warms up and a **glowing tail** is thrown out. The tail is made up of the dust and rocks freed from the ice.
- Compared to the near circular orbits of the planets, the orbit of a comet is very **elliptical**. Most comets pass inside the orbit of Mercury and well beyond the orbit of Pluto.
- Solar winds blow the dust into the comet's tail so the tail always points away from the Sun.

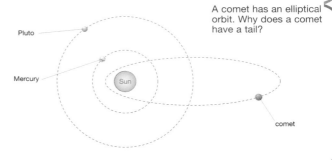

A comet has an elliptical orbit. Why does a comet have a tail?

NEOs

- Some comets and asteroids have orbits that pass close to the orbit of the Earth. These are known as near-Earth-objects (**NEOs**) and some can be seen with a telescope. A small change in their direction could bring them on a collision course with the Earth.
- Scientists are constantly monitoring and plotting the paths of comets and other NEOs.

Top Tip! The longer we study and plot the path of a NEO, the more accurate our prediction about its future movement and the risk of collision will be.

Questions

Grades G-E
1 What could happen if a large asteroid hit Earth?

Grades D-C
2 What's an asteroid?

Grades G-E
3 What forms the tail of a comet?

Grades D-C
4 Why is it important to constantly monitor the paths of comets and NEOs?

The Big Bang

The Universe

- Fifteen billion years ago, all of the matter in the Universe was in a single point. It was incredibly hot – millions upon millions of degrees Celsius.

- The Universe suddenly exploded; scientists call this the **Big Bang**. It expanded rapidly and cooled down. Within three seconds, electrons, protons, neutrons and the elements **hydrogen** and **helium** had formed.

- Almost all of the **galaxies** in the Universe are **moving away** from each other. The furthest galaxies are moving fastest. The Universe is expanding all the time.

- **Microwave signals** are constantly reaching Earth from all parts of the Universe.

How does this model show that the Universe is expanding?

Stars

- **New stars** are being formed all the time. They start as a swirling cloud of gas and dust.

- Eventually all stars will die and some stars become **black holes**. Light can't escape from a black hole.

- A medium-sized star, like the Sun, becomes a **red giant**: the core contracts, the outer part expands and cools, and it changes colour from yellow to red. During this phase, gas shells, called **planetary nebula**, are thrown out.

- The core becomes a **white dwarf** shining very brightly, but eventually cools to become a **black dwarf**.

- Large stars become **red supergiants** as the core contracts and the outer part expands.

- The core suddenly collapses to form a **neutron star** and there's an explosion called a **supernova**. Remnants from a supernova can merge to form a **new star**. The dense core of the neutron star continues to collapse until it becomes so dense it forms a **black hole**.

A star is born!

Questions

Grades G-E

1. What were the first elements formed after the Big Bang?

Grades D-C

2. Which galaxies in the Universe are moving fastest?

Grades G-E

3. Why can't we see a black hole?

Grades D-C

4. Our Sun is about 4.6 billion years old. Patrick says it may have been a star before that. Explain how this is possible.

P2 Summary

Energy sources

- **Kinetic energy** from moving air turns the blades on a wind turbine to produce electricity.
- The **Sun** is a stable energy source. It **transfers energy** as light and heat to Earth.
- **Photocells** use the Sun's light to produce electricity.
- Passive solar heating uses glass to help keep buildings warm.

Electricity generation

- A **dynamo** produces electricity when coils of wire rotate inside a magnetic field. The size of the current depends on:
  – the number of turns
  – the strength of the field
  – the speed of rotation.
- **Transformers** change the size of the voltage and current. The National Grid transmits electricity around the country at **high voltage** and **low current**. This reduces energy loss.
- **Nuclear fuels** are radioactive. The radiation produced can cause cancer. Waste products remain radioactive for a long time.
- **Fossil fuels** and **biomass** are burned to produce heat. Nuclear fuels release energy as heat. Water is heated to produce steam:
  – the steam drives turbines
  – turbines turn generators
  – generators produce electricity.
- The main forms of **ionising radiation** are:
  – **alpha**
  – **beta**
  – **gamma**.
  Their uses depend on their penetrative and ionisation properties.

The Earth's field

- The **Earth** is surrounded by a **magnetic field** similar in shape to that of a coil of wire. The core of the Earth contains molten iron.
- **Solar flares** are the result of clouds of charged particles being emitted from the Sun at high speed disturbing the magnetic field around Earth.
- When two planets collide, a new planet and a moon may be formed.

The Earth and Universe

- **Planets, asteroids** and **comets** orbit the Sun in our **Solar System**. The Universe consists of many galaxies.
- Most **asteroids** are between Mars and Jupiter but some pass closer to Earth. They are constantly being monitored. An asteroid strike could cause climate change and species extinction.
- Scientists believe that the Universe started with a **Big Bang**. **Stars** have a **finite life** depending on their size.
- The **Universe** is explored by telescopes on Earth and in space. Large distances mean that it takes a long time for information to be received and inter-galactic travel is unlikely.

Speed

Measuring speed

- **Speed** is a measure of how fast something is going.
- To determine speed you need to measure **distance** and **time**. Faster objects cover more distance in a given time.
- The unit of speed is **metres per second** (m/s) or **kilometres per hour** (km/h). So, if a car is travelling at 80 km/h it goes a distance of 80 km in a time of 1 hour.
- In everyday situations a **tape measure** or **trundle wheel** and **stopwatch** can be used to measure speed. On roads, a speed camera takes a photograph as a car passes. A second photograph is taken 0.5 seconds later. There are white lines painted on the road 1.5 m apart which show how far the car travels in 0.5 seconds.

A car passing over six lines travels:
$1.5 \times 6 = 9$ m in 0.5 seconds
This means the average speed of the car was 18 m/s or 65 km/h.

Top Tip! Remember! 1 km = 1000 m, 1 hour = 60 × 60 = 3600 seconds.

- The formula for speed is: average speed $= \dfrac{\text{distance}}{\text{time}} = \dfrac{d}{t}$

We write 'average speed' because the speed of a car changes during a journey.

An aircraft travels 1800 km in 2 hours.
Average speed $= \dfrac{1800}{2} = 900$ km/h $= 900 \times \dfrac{1000}{3600} = 250$ m/s.

- Increasing the speed means increasing the distance travelled in the same time. Increasing the speed reduces the time needed to cover the same distance.

Distance–time graphs

- Drawing a graph of distance against time shows how the distance moved by an object from its starting point changes over time.
- In graph **a**, the distance doesn't change over time, so the car is stationary. In graph **b**, the distance travelled by the car increases at a steady rate. It's travelling at a constant speed.

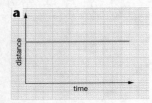
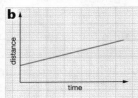
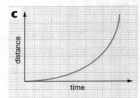

- The **gradient** of a distance–time graph tells you about the speed of the object. A higher speed means a steeper gradient.
- In graph **b**, the distance travelled by the object each second is the same. The gradient is constant, so the speed is constant.
- In graph **c**, the distance travelled by the object each second increases as the time increases. The gradient increases, so there is an increase in the speed of the object.

Questions

Grades G-E
1. What is the unit of speed?

Grades D-C
2. A car travels 600 m in 20 seconds. What is its average speed?

Grades G-E
3. Sketch distance–time graphs for **a** a stationary train **b** a train travelling at a steady speed.

Grades D-C
4. What can you say about the gradient of a distance–time graph for a car journey if **a** the car is travelling at a steady speed **b** the speed of the car is decreasing?

Changing speed

Speed–time graphs

- A speed–time graph shows how the **speed** of an object, for example a car, changes with **time**.
- The **gradient** (slope) of a line tells us how the speed is changing. In graph **a**, the line is horizontal so the speed is constant. In graph **b**, the line has a positive gradient so the speed is increasing. In graph **c**, the line has a negative gradient so the speed is decreasing.

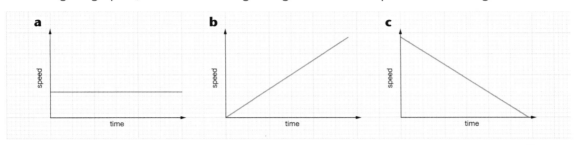

- If the speed is increasing, the object is **accelerating**. If the speed is decreasing, the object is **decelerating**.
- The area under a speed–time graph is equal to the **distance** travelled.
  – The speed of car B in the graph is increasing more rapidly than the speed of car A, so car B is travelling further than car A in the same time.
  – The area under line B is greater than the area under line A for the same time.
  – The speed of car D is decreasing more rapidly than the speed of car C, so car D isn't travelling as far as car C in the same time.
  – The area under line D is smaller than the area under line C for the same time.

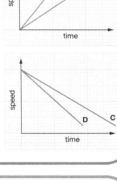

Top Tip! Don't confuse distance–time and speed–time graphs. Always look at the axes carefully.

Acceleration

- A change of speed is called **acceleration**.
- Acceleration is measured in **metres per second squared** (m/s^2).
- In graph **a**, the car has a constant acceleration of 5 m/s^2. In graph **b**, the car has a constant deceleration of 5 m/s^2.

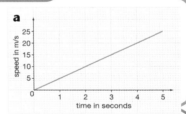
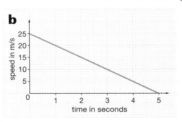

- The formula for measuring acceleration is:

$$\text{acceleration} = \frac{\text{change in speed}}{\text{time taken}}$$

- A negative acceleration shows the car is decelerating.

A new car boasts a rapid acceleration of 0 to 108 km/h in 6 seconds.

A speed of 108 km/h is $\frac{108 \times 1000}{60 \times 60} = 30$ m/s

Acceleration = $\frac{\text{change in speed}}{\text{time taken}} = \frac{(30 - 0)}{6} = 5$ m/s^2

This means the speed of the car increases by 5 m/s every second.

Top Tip! An object at a constant acceleration is different to an object at a constant speed.

Questions

Grades G-E
1. Sketch a speed–time graph for a car that speeds up and then travels at a constant speed.

Grades D-C
2. What can be found from the area under a speed–time graph?

3. The speed of a car changes from 20 m/s to 10 m/s. Is it accelerating or decelerating?

4. Find the acceleration of a cat that goes from 0 to 5 m/s in 2 seconds when chasing a mouse.

Forces and motion

What do forces do?

- To **accelerate** in a car, the driver presses on the accelerator pedal.
  - This increases the pull of the engine (forward **force**).
  - If the pedal is pressed down further, the pull of the engine is greater.
  - The **acceleration** increases.

for the same forward force:	for the same mass:	for the same acceleration:
more mass has less acceleration	more forward force causes more acceleration	a large mass needs a large forward force
less mass has more acceleration	less forward force causes less acceleration	a small mass needs a small forward force

Force, mass and acceleration

- If the forces acting on an object are balanced, it's at rest or has a constant speed. If the forces acting on an object are unbalanced, it speeds up or slows down.
- The unit of force is the **Newton** (N).
- $F = ma$, where F = unbalanced force in N, m = mass in kg and a = acceleration in m/s^2.

Marie pulls a sledge of mass 5 kg with an acceleration of $2\ m/s^2$ in the snow. The force needed to do this is: $F = ma$ $F = 5 \times 2 = 10\ N$

Top Tip! Remember the correct units when using $F = ma$; F in N, m in kg, a in m/s^2.

Car safety

- A car driver cannot stop a car immediately. It takes the driver time to react to danger.
  - **Thinking distance** is the distance travelled between a driver seeing a danger and taking action to avoid it.
  - **Braking distance** is the distance travelled before a car comes to a stop after the brakes have been applied.
- The formula for **stopping distance** is:
  stopping distance = thinking distance + braking distance

- **Thinking time**, and therefore thinking distance, may increase if a driver is:
  - tired
  - under the influence of alcohol or other drugs
  - distracted or lacks concentration.
- For safe driving, it is important to:
  - keep an appropriate distance from the car in front
  - have different speed limits for different types of road
  - slow down when road conditions are poor.

Top Tip! For an alert driver, thinking time (or **reaction time**) is about 0.7 seconds.

Questions

Grades G-E
1. Two cars, A and B, have the same engine. Car A has a larger mass than car B. Which car has the greater acceleration?

Grades D-C
2. Calculate the unbalanced force needed to give a car of mass 1200 kg an acceleration of $5\ m/s^2$.

Grades G-E
3. What do we call the distance travelled by a car in stopping after the brakes have been applied?

Grades D-C
4. Tim's reaction time is 0.7 seconds. Calculate his thinking distance when travelling at 20 m/s.

Work and power

Work

(Grades G–E)

- **Work** is done when a **force** moves. For example, when a person climbs stairs, the force moved is their weight.
- The amount of work done depends on:
  – the **size** of the force acting on an object
  – the **distance** the object is moved.
- **Energy** is needed to do work. The more work is done, the more energy is needed.
- Work and energy are measured in **joules** (J).

(Grades D–C)

- **Work** is done when a force moves in the direction in which the force acts.
- The formula for work done is:
  work done = force × distance moved (in the direction of the force)
  If a person weighs 700 N, the work he does against gravity when he jumps 80 cm is:
  work done = force × distance moved = 700 × 0.8 = 560 J

Power

(Grades G–E)

- **Power** is the rate at which work is done.
- If two lifts, one old and one new, do the same amount of work but the new one does it more quickly, the new lift has a greater power.
- Power is measured in **watts** (W). A large amount of power is measured in **kilowatts** (kW). 1 kW = 1000 W.

(Grades D–C)

- The formula for power is:
  $$\text{power} = \frac{\text{work done}}{\text{time taken}}$$
- A person's power is greater when they run than when they walk.

Top Tip! Always check the units when doing calculations on work, power and energy.

Fuel

(Grades G–E)

- Some cars are more powerful than others. They travel faster and cover the same distance in a shorter time and require more fuel. They have greater **fuel consumption**.
- Fuel is expensive and a car with high fuel consumption is expensive to run.

(Grades D–C)

- Fuel pollutes the environment.
  – Car exhaust gases, especially carbon dioxide, are harmful.
  – Carbon dioxide is also a major source of **greenhouse gases**, which contribute to climate change.

Questions

(Grades G-E)
1. Sam lifts an 8 N weight a height of 3 m. Ali lifts a 10 N weight a height of 4 m. Who does more work?

(Grades D-C)
2. How much work does Lee do when he pushes a car 20 m with a force of 80 N?

(Grades G-E)
3. Jan and Meera both run up a flight of stairs in 8 seconds. Meera is more powerful than Jan. Explain.

(Grades D-C)
4. Find the power of a crane that can lift a load of 500 N through a height of 12 m in 15 seconds.

Energy on the move

Kinetic energy

- Moving objects have **kinetic energy**. Different fuels can be used to gain kinetic energy.
  - The fuel for all animals, including humans, is their food.
  - The fuel for a wind turbine is moving air (wind).
  - The fuel for a car is **petrol** or **diesel** oil.

- **Kinetic energy** increases with:
  - increasing **mass**
  - increasing **speed**.

Fuel

- **Petrol** and **diesel oil** are **fossil fuels** made from **oil**. Petrol is more **refined** than diesel oil.

- Petrol cars and diesel cars need different engines. The same amount of diesel oil contains more **energy** than petrol. A diesel engine is more **efficient** than a petrol engine.

- Some cars use more petrol or diesel oil than others and:
  - cause more **pollution**
  - cost more to run
  - decrease supplies of **non-renewable** fossil fuels.

- **Fuel consumption** data are based on ideal road conditions for a car driven at a steady speed in urban and non-urban conditions.

Top Tip! Make sure you can interpret fuel consumption tables.

car	fuel	engine in litres	miles per gallon (mpg)	
			urban	non-urban
Renault Megane	petrol	2.0	25	32
Land-Rover	petrol	4.2	14	24

Electrically powered cars

- **Electric cars** are battery driven or solar-powered.

- Exhaust fumes from petrol-fuelled and diesel-fuelled cars cause serious pollution in towns and cities.

- Battery-driven cars do not pollute the local environment, but their batteries need to be recharged. Recharging uses electricity from a power station. Power stations pollute the local atmosphere and cause acid rain.

This hybrid electric car has solar panels on its roof that convert sunlight into additional power to supplement its battery.

Questions

Grades G-E
1 Name the energy source for **a** a bird **b** a barbecue **c** a hydroelectric power station.

Grades D-C
2 Suggest why a car uses more fuel per kilometre when carrying a heavy load.

Grades G-E
3 A petrol-driven car does 40 mpg (miles per gallon) and a similar diesel-driven car does 56 mpg. Which car is more efficient?

Grades D-C
4 Electric milk floats have been used for many years. Suggest why more people don't use electric cars.

Crumple zones

Car safety

Grades G–E

- A moving car has **kinetic energy**. If a car is involved in a collision it has to lose kinetic energy very quickly.

- Modern cars have safety features that **absorb energy** when a vehicle stops suddenly. These are:
  - **brakes** that get hot
  - **crumple zones** that change shape
  - **seat belts** that stretch a little
  - **air bags** that inflate and squash

- Safety features can be **active** or **passive**. These are shown in the table.

active safety features	passive safety features
ABS (anti-lock braking system) brakes	electric windows
	cruise control
traction control	paddle shift controls
safety cage	adjustable seating

- All safety features must be kept in good repair.
  - Seat belts must be replaced after a crash in case the fabric has been overstretched.
  - The safety cage must be examined for possible damage after a crash.
  - Seat fixings should be checked frequently to make sure they are secure.

Grades D–C

- On impact:
  - **crumple zones** at the front and rear of the car absorb some of its energy by changing shape or 'crumpling'
  - a **seat belt** stretches a little so that some of the person's kinetic energy is converted to elastic energy
  - an **air bag** absorbs some of the person's kinetic energy by squashing up around them.

An air bag and seat belt in action during a car accident. What parts of the man are protected?

- All these safety features:
  - absorb energy
  - change shape
  - reduce injuries.

- Active safety features include:
  - **ABS brakes** which give stability and maintain steering during hard braking. The driver gets the maximum braking force without skidding and can still steer the car.
  - **Traction control** which stops the wheels on a vehicle from spinning during rapid acceleration.
  - A car **safety cage** which is a **rigid** frame that prevents the car from collapsing and crushing the occupants in a roll-over crash.
  - Crumple zones at front and rear which keep damage away from the internal safety cage.

Top Tip!
ABS brakes don't stop a car more quickly. They give improved control and prevent skidding.

- Passive safety features help the driver to concentrate on the road.

Questions

Grades G-E

1 Suggest two forms of energy that a car's kinetic energy may be changed to in an accident.
2 Explain how electric windows can help to prevent car accidents.

Grades D-C

3 Which two car safety features protect a driver's chest in an accident?
4 Explain why cruise control is a passive safety feature.

Falling safely

Falling objects

- A falling object:
  - gets faster as it falls
  - is pulled towards the centre of Earth by a force called **weight** which is caused by the force of **gravity**.

> **Top Tip!** The net force acting on a parachutist = weight − air resistance.

- Objects that have a large **area of cross-section** fall more slowly. For example, if a ball and a feather are dropped, the ball hits the ground first. This slowing down **force** is called **air resistance** or **drag**.

- All objects fall with the same **acceleration** as long as the effect of **air resistance** is very small.

- The size of the air resistance force on a falling object depends on:
  - its cross-sectional area – the larger the area the greater the air resistance
  - its speed – the faster it falls the greater the air resistance.

- Air resistance has a significant effect on motion only when it is large compared to the weight of the falling object.

- The speed of a **free-fall** parachutist changes as he falls to Earth.
  - In picture 1, the weight of the parachutist is greater than air resistance. He accelerates.
  - In picture 2, the weight of the parachutist and air resistance are equal. The parachutist has reached **terminal speed** because the forces acting on him are **balanced**.
  - In picture 3, the air resistance is larger than the weight of the parachutist. He slows down and air resistance decreases.
  - In picture 4, the air resistance and weight of the parachutist are the same. He reaches a new, slower terminal speed.

1 600 N

2 600 N / 600 N

1000 N
3 600 N

600 N
4 600 N

- Truly free-falling objects:
  - don't experience air resistance
  - accelerate downwards at the same rate irrespective of mass or shape.

- Examples of truly free-falling objects are:
  - objects falling above the Earth's atmosphere
  - objects falling on the Moon   – the Moon itself.

> **Top Tip!** On the Moon and in outer Space there is no drag because there is no atmosphere.

Friction

- All **friction** forces slow an object down.

- Friction forces on vehicles can be reduced by **streamlining** their shapes by:
  - shaping car roof boxes
  - making cars wedge-like in shape
  - angling lorry deflectors.

Streamlining reduces drag and increases the top speed.

Questions

Grades G-E
1. Why does a feather fall more slowly than a ball?

Grades D-C
2. Name the two forces acting on a parachutist. What can you say about these forces **a** at the start of the descent **b** when terminal speed is reached **c** when the parachute opens.

Grades G-E
3. What effect does a streamlined shape have on the motion of a car?

Grades D-C
4. Describe how air resistance changes with the speed of a fallling object.

The energy of games and theme rides

Gravitational potential energy

- An object held above the ground has **gravitational potential energy**.
- The amount of gravitational potential energy an object has depends on:
  – its **mass**   – its **height** above the ground.

Energy transfers

- A bouncing ball converts gravitational potential energy to kinetic energy and back to gravitational potential energy. It does not return to its original height because energy is transferred to other forms such as **thermal energy** and **sound energy**.
- In the diagram, the gravitational potential energy at D is less than at A.

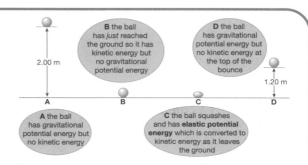

The stages of energy transfer when a ball is dropped.

A the ball has gravitational potential energy but no kinetic energy

B the ball has *just* reached the ground so it has kinetic energy but no gravitational potential energy

C the ball squashes and has **elastic potential energy** which is converted to kinetic energy as it leaves the ground

D the ball has gravitational potential energy but no kinetic energy at the top of the bounce

- A water-powered **funicular railway** has a carriage that takes on water at the top of the hill giving it extra gravitational potential energy.
  – As the carriage travels down the hillside it transfers gravitational potential energy to **kinetic energy**.
  – At the same time, it pulls up another carriage with an empty water tank on a parallel rail.
- A moving object has **kinetic energy**.
- The amount of kinetic energy an object has depends on:
  – its **mass**   – its **speed**.

A funicular railway.

How a roller coaster works

- A roller coaster uses a motor to haul a train up in the air. The riders at the top of a roller coaster ride have a lot of gravitational potential energy.
- When the train is released it converts **gravitational potential energy** to **kinetic energy** as it falls. This is shown by the formula:
  loss of gravitational potential energy (PE) = gain in kinetic energy (KE)
- Each peak is lower than the one before because some energy is transferred to heat and sound due to friction and air resistance. This is shown by the formula:
  PE at top = KE at bottom + energy transferred (to heat and sound) due to friction
- kinetic energy = $\frac{1}{2}mv^2$
  – If **speed doubles, KE quadruples** ($KE \propto v^2$).
  – If **mass doubles, KE doubles** ($KE \propto m$).

Top Tip!
Remember! Energy is always conserved.

Questions

Grades G-E
1 Alfie, on the 3rd floor, and Jo, on the 8th floor, each hold similar balls out of a window. Whose ball has the greater gravitational potential energy?

Grades D-C
2 Describe the energy changes for a girl on a swing.

Grades G-E
3 Why does a funicular railway pull up a carriage with an empty water tank?

Grades D-C
4 Why does a roller coaster travel more slowly at the top than at the bottom of each hill?

P3 Summary

P3 FORCES FOR TRANSPORT

Distance–time graphs

The object **isn't moving**.

The object has a **constant speed**.

Speed and acceleration

$$\text{speed} = \frac{\text{distance}}{\text{time}}$$

$$\text{acceleration} = \frac{\text{change in speed}}{\text{time}}$$

Speed–time graphs

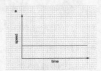

The speed is **constant**.

The speed is **increasing**.

Forces for transport

Force, mass and acceleration are linked by the equation: $F = ma$

Fossil fuels – for example, petrol and diesel – are the main fuels used in road transport. Some cars use more petrol or diesel than others so cause more pollution. Diesel engines usually have better **fuel consumption** figures than petrol engines.

Forces can make things speed up or slow down.

When a car stops the:
total stopping distance = thinking distance + braking distance
The higher the speed of the car, the greater the distance it travels in stopping.

Work, energy and power

Work is done when a force moves an object.
$W = F \times d$
Work is measured in joules (J).

Roller coasters use **gravitational potential energy** as the source of movement.

Power is a measure of how quickly work is done.
$P = \frac{W}{t}$
Power is measured in watts (W).

Modern cars have lots of safety features that absorb energy when the cars stop:
– crumple zones
– seatbelts
– air bags.
Cars also have:
– **active safety features** (e.g. ABS brakes)
– **passive safety features** (e.g. cruise control).

Moving objects possess **kinetic energy**.
– The faster they travel, the more kinetic energy they possess.
– The greater the mass of an object, the more kinetic energy it possesses.
kinetic energy (KE) = $\frac{1}{2}mv^2$

Energy is needed to do work. Energy is measured in joules (J).

Falling safely

Terminal speed is the maximum speed reached by a falling object. It happens when the forces acting on the object are balanced.

Falling objects get faster as they fall. They're pulled towards the centre of the Earth by their weight (**gravity**).

A parachute provides a large upward force, bigger than the weight of the parachutist. This slows him down. As he slows down the air resistance gets less until it equals his weight. He then falls at a terminal speed and lands safely.

Sparks!

Grades

Insulating materials
G–E
- Metals are good electrical **conductors**. They allow electric charges to move through them.
- Materials such as wood, glass and polythene are **insulators**. They do *not* allow electric charges to pass through them.

Positive and negative charges
G–E
- Charge can build up on an insulator. An insulator can be charged by friction.
- There are two kinds of electric charge, **positive** and **negative**. When rubbed with a duster:
  - acetate and perspex become positively charged
  - polythene becomes negatively charged.

D–C
- An **atom** consists of a small positively charged nucleus surrounded by an equal number of negatively charged electrons.
- In a stable, neutral atom, there are the same amounts of positive and negative charges.
- All **electrostatic** effects are due to the movement of electrons.
- The law of electric charge states that: like charges repel, unlike charges attract.

Top Tip! It's only electrons that move in an atom.

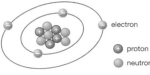

The charges in a neutral atom balance. The atom has four electrons. How many protons does it have?

Electric shocks
G–E
- A person gets an electric shock if they become charged and then become earthed.
- A person can become charged if they walk on a nylon-carpeted or vinyl floor because:
  - the floor is an insulator
  - they become charged as they walk due to friction.
- The person can become earthed by touching water pipes or even another person.
- Synthetic clothing can also cause a person to become charged.

Dangers of static electricity
D–C
- When inflammable gases or vapours are present, or there is a high concentration of oxygen, a spark from static electricity could ignite the gases or vapours and cause an explosion.
- If a person touches something at a high **voltage**, large amounts of electric charge may flow through their body to earth.
- **Current** is the rate of flow of charge. Even small currents can be fatal.
- Static electricity can be a nuisance but not dangerous.
  - Dust and dirt are attracted to insulators, such as television screens.
  - Clothes made from synthetic materials often 'cling' to each other and to the body.

P4 RADIATION FOR LIFE

Questions

Grades G-E
1. Which of these materials are electrical insulators?
   a copper **b** paper **c** polythene **d** tin.

Grades D-C
2. What happens when **a** two acetate rods are brought near each other? **b** a polythene and an acetate rod are brought near each other?

Grades G-E
3. Suggest why synthetic clothes may become charged.

Grades D-C
4. Why must a mobile phone *not* be used on a petrol station forecourt?

Uses of electrostatics

Uses of static electricity

- A paint sprayer **charges** paint droplets to give an even coverage.
- A photocopier and laser printer use charged particles to produce an image.
- Charged plates inside factory chimneys are used to remove dust particles from smoke.
- A **defibrillator** delivers a controlled electric **shock** through a patient's chest to restart their heart.

Defibrillators

- Defibrillation is a procedure to restore a regular **heart rhythm** by delivering an electric shock through the chest wall to the heart.
  - Two **paddles** are charged from a high voltage supply.
  - They are then placed firmly on the patient's chest to ensure good electrical contact.
  - Electric charge is passed through the patient to make the heart contract.
  - Great care is taken to ensure that the operator does not receive an electric shock.

Paint sprayers

- Static electricity is used in paint spraying.
  - The **spray gun** is charged.
  - All the paint particles become charged with the same charge.
  - Like charges **repel**, so the paint particles spread out giving a fine spray.
  - The object to be painted is given the *opposite* charge to the paint.
  - Opposite charges **attract**, so the paint is attracted to the object and sticks to it.

- The advantages of electrostatic paint sprayers are:
  - less paint is wasted
  - the object gets an even coat of paint
  - paint gets into awkward corners well.

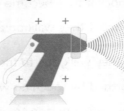
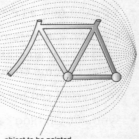

How an electrostatic paint sprayer works.

nozzle is charged up positively

object to be painted is negatively charged

Dust precipitators

- A dust precipitator removes harmful particles from the chimneys of factories and power stations that pollute the atmosphere.
  - A metal grid (or wires) is placed in the chimney.
  - The grid is connected to a high-voltage supply.
  - Dust particles are attracted to the metal grid.
  - The dust particles stick together to form larger particles.
  - When these particles get big enough they fall back down the chimney.

Questions

Grades D-C

1. Suggest why the paddles of a defibrillator must be placed firmly on the bare chest of the patient.
2. If the paint from an electrostatic paint spray is positively charged, what charge is given to the object to be painted?
3. Why do the wires in an electrostatic dust precipitator need to be at a high voltage?

Safe electricals

Electric circuits

- A closed loop, with no gaps, is required for a **circuit** to work.

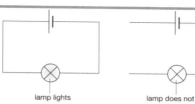

lamp lights — lamp does not light

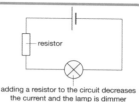

adding a resistor to the circuit decreases the current and the lamp is dimmer

- An electric **current** is a flow of electric **charge**. Charge is carried by negatively charged **electrons**. The electrons flow in the *opposite* direction to the conventional current.
- The current is measured in **amperes** (A) using an **ammeter** connected in **series**.

Resistance and resistors

- A **resistor** is added to a circuit to change the amount of **current** in it.
- A **variable resistor** (or **rheostat**) changes the resistance.
- The **potential difference** (pd) between two points in a circuit is the difference in **voltage** between the two points.

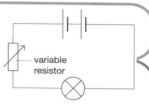

This circuit has a variable resistor and acts as a dimmer switch.

- Potential difference is measured in **volts** (V) using a **voltmeter** connected in **parallel**.
  – For a fixed resistor, as the potential difference across it **increases**, the **current increases**.
  – For a fixed power supply, as the **resistance increases**, the **current decreases**.
- The formula for resistance is: $\text{resistance} = \dfrac{\text{potential difference}}{\text{current}}$ $R = \dfrac{V}{I}$
  – Resistance is measured in **ohms** (Ω).

  If the pd across a lamp is 5.0 V when the current is 0.2 A, the resistance of the lamp, $R = \dfrac{V}{I} = \dfrac{5.0}{0.2} = 25\ \Omega$

Top Tip! Always remember to include the correct unit.

Live, neutral and earth wires

- The cable used to connect an appliance to the mains has three wires inside it. **Live**, **neutral** and **earth wires** can be seen in a plug. An earthed conductor can't become live.

live wire	neutral wire	earth wire
brown	blue	green/yellow striped
carries a high voltage around the house.	completes the circuit, providing a return path for the current	safety wire connected to the metal case of an appliance to prevent it becoming live

Wiring inside a plug.

Fuses and insulation

a Earth symbol
b double-insulation symbol.

- A **fuse** breaks a circuit if a fault occurs. A **circuit breaker** is a re-settable fuse.
- Double-insulated appliances don't need an earth wire. The case cannot become live.
- A fuse contains a length of wire which melts, breaking the circuit, if the current becomes too large. Fuse values of 5 A and 13 A are commonly available for use in three-pin plugs.

Questions

Grades G-E
1 What happens to the current in a circuit if the resistance increases?

Grades D-C
2 The pd across a resistor is 4.0 V when the current through it is 0.5 A. What is its resistance?

Grades G-E
3 Which mains wire only carries a current when a fault occurs?

Grades D-C
4 Explain how a fuse protects an appliance.

Ultrasound

Longitudinal waves

- All **sound**, including **ultrasound**, is produced by **vibrating** particles that form a **longitudinal wave**.
  - The vibrations of the particles are in the **same** direction as the wave.
  - This sets up a pressure wave with **compressions** and **rarefactions**.
  - The pressure wave makes a person's eardrums vibrate and signals are sent to the brain.
- The features of a wave are:
  - **Amplitude** (A), which is the maximum distance a particle moves from its normal position.
  - **Wavelength** (λ), which is the distance occupied by one complete wave.
  - **Frequency** (f), which is the number of complete waves in a second. Frequency is measured in **hertz** (Hz). A frequency of 50 Hz means there are 50 complete waves in a second.

Top Tip! With sound waves, **frequency** is linked to **pitch** and **amplitude** to **loudness**.

A sound wave showing rarefactions, compressions and wavelength.

- The features of longitudinal sound waves are:
  - They can't travel through a vacuum. The denser the medium, the faster a wave travels.
  - The higher the frequency or pitch, the smaller the wavelength.
  - The louder the sound, or the more powerful the ultrasound, the more energy is carried by the wave.
- **Ultrasound** is sound of a higher **frequency** than humans can hear.
- Ultrasound travels as a pressure wave – compressions and rarefactions.

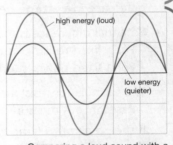
Comparing a loud sound with a quieter sound. How does the amplitude change?

Uses of ultrasound

- **Ultrasound** can be used:
  - in body scans
  - to break down kidney stones
  - to measure the speed of blood flow in a vein or artery when a blockage is suspected.
- A body scan:
  - allows doctors to 'see' inside a patient without surgery
  - can investigate heart and liver problems
  - can check the condition of a foetus
  - can look for tumours in the body.
- A **pulse** of ultrasound is sent into the body:
  - at each boundary between different tissues or organs, some ultrasound is **reflected** and the rest is transmitted
  - the returning **echoes** are recorded and used to build up an image of the internal structure.
- To break down kidney stones:
  - a high-powered ultrasound beam is directed at the kidney stones
  - the ultrasound energy breaks the stones down into smaller pieces
  - the tiny pieces are then excreted from the body in the normal way.

Questions

Grades G-E
1. Middle C has a frequency of 256 Hz. How many vibrations occur every second?

Grades D-C
2. Body fat is denser than air. In which medium will ultrasound travel faster?

Grades G-E
3. What is the advantage of using ultrasound to break down kidney stones?

Grades D-C
4. An ultrasound pulse travels 20 cm further when it is reflected from one side of the head of a foetus compared with the other side. How big is the head?

Treatment

Radiation

- X-rays and gamma rays are both **electromagnetic waves** with very short wavelengths. They are very penetrating so can pass into the body to treat internal organs.

- **X-rays** and **gamma (γ) rays** are used in medicine:
  – for diagnosis (finding out what is wrong)
  – for therapy (treatment).

- **Gamma radiation** is emitted from the **nucleus** of an **atom**.

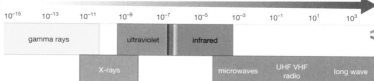

The electromagnetic spectrum.

- Radiation emitted from the nucleus of an unstable atom can be **alpha** (α), **beta** (β) or **gamma** (γ).
  – **Alpha radiation** is absorbed by the skin, so is of no use for diagnosis or therapy.
  – **Beta radiation** passes through skin but not bone. Its medical applications are limited but it's used, for example, to treat the eyes.
  – **Gamma radiation** is very penetrating and is used in medicine. **Cobalt-60** is a gamma-emitting radioactive material that is widely used to treat cancers.

- Gamma rays and X-rays have similar wavelengths but they're produced in different ways.

- When nuclear radiation passes through a material it causes **ionisation**. Ionising radiation damages living cells, increasing the risk of **cancer**.

- Cancer cells within the body can be destroyed by exposing the affected area to large amounts of radiation. This is called **radiotherapy**.

Gamma radiation

- Gamma radiation is used to treat **cancer**.
  – Large doses of high-energy radiation can be used in place of surgery. However, it's more common to use radiation after surgery, to make sure *all* the cancerous cells are removed or destroyed.
  – If any cancerous cells are left behind, they can multiply and cause **secondary cancers** at different sites in the body.

- Gamma radiation kills bacteria. It's used to **sterilise** hospital equipment to prevent the spread of disease. Each item is put in a sealed bag and exposed to intense gamma radiation.

- A **radiographer** carries out procedures using X-rays and nuclear radiation. Sterilising medical equipment.

Tracers

- A radioactive **tracer** is used to investigate inside a patient's body without surgery. Beta or gamma emitters are used.
  – Technetium-99m is a commonly used tracer. It emits only gamma radiation.
  – Iodine-123 emits gamma radiation and is used as a tracer to investigate the thyroid gland.
  – X-rays cannot be used as tracers as they're produced in an X-ray machine.

Questions

Grades G-E
1 Why do X-rays and gamma rays have many medical uses?

Grades D-C
2 What is 'radiotherapy'?

Grades G-E
3 Why is gamma radiation used to sterilise equipment?

Grades D-C
4 Why is iodine-123 used as a tracer in medicine?

What is radioactivity?

Radioactivity and radioactive decay

- A Geiger-Muller tube and ratemeter (together commonly called a Geiger counter) are used to detect the rate of **decay** of a radioactive substance.
  - Each 'click' or extra number on the display screen represents the decay of one **nucleus**.
  - A decaying nucleus emits **radiation**.

- The **activity** is measured by the average number of nuclei that decay every second.
  - This is also called the **count rate**.
  - Activity is measured in counts/second or **becquerels** (**Bq**).
  - The formula for activity is:

  $$\text{activity} = \frac{\text{number of nuclei which decay}}{\text{time taken in seconds}}$$

- The activity of a radioactive substance **decreases** with time – the count rate falls.

- Radioactive substances decay naturally, giving out **alpha**, **beta** and/or **gamma** radiation.

- Radioactive decay is a **random** process; it isn't possible to predict exactly when a nucleus will decay.
  - There are so many atoms in even the smallest amount of **radioisotope** that the average count rate will always be about the same.

- Radioisotopes have **unstable nuclei**. Their nuclear particles aren't held together strongly enough.

Top Tip! Remember! Alpha, beta and gamma radiation are not radioactive; it's the source that emits them that is radioactive.

What are alpha and beta particles?

An **alpha (α) particle**
- is positively charged
- has quite a large mass
- is a helium nucleus
- consists of two **protons** and two **neutrons**.

A **beta (β) particle**
- is negatively charged
- has a *very* small mass
- travels *very* fast (at about one-tenth of the speed of light)
- is a fast-moving **electron**.

Top Tip! Know the difference between alpha, beta and gamma radiation.

- When an **alpha** or a **beta** particle is emitted from the nucleus of an atom, the remaining nucleus is a **different** element. But **gamma** radiation it is not a particle, so does *not* change the composition of the nucleus; it remains the **same** element.

Alpha particles emitted from radiation

Questions

Grades G-E

1. Ali records a count of 6000 in 10 seconds from a radioactive source. What is its activity?

Grades D-C

2. Three successive measurements of the activity of a radioactive source are 510 Bq, 495 Bq and 523 Bq. Why are they different?
3. What is **a** an alpha particle **b** a beta particle?
4. Why does the emission of a gamma ray not change the composition of the nucleus?

Uses of radioisotopes

Background radiation

- **Background radiation** is ionising radiation that is always present in the environment.
  - It varies from place to place and from day to day.
  - The level of background radiation is low and doesn't cause harm.

- **Background radiation** is caused by:
  - radioactive substances present in **rocks** (especially **granite**) and soil
  - **cosmic rays** from Space.

Top Tip!
Remember to allow for background radiation when measuring the activity of a source of radiation.

Tracers

- **Radioisotopes** are used as **tracers** in industry and in hospitals.

- In addition to medical applications (see page 35), **tracers** are used to:
  - track the dispersal of waste materials
  - find leaks or blockages in underground pipes
  - track the route of underground pipes.

Smoke alarms

- One type of **smoke detector** uses a source of **alpha particles** to detect smoke. It's sensitive to low levels of smoke.

- Many **smoke detectors** contain a radioisotope such as **americium-241** which emits alpha particles. It has a long half life so decays slowly.
  - The alpha particles ionise some of the oxygen and nitrogen atoms in the air.
  - The positive ions and negative electrons move towards the negative and positive plates respectively (opposite charges attract).
  - This creates a tiny current which is detected by electronic circuitry in the smoke alarm.
  - If smoke particles enter, they attach themselves to the ions, neutralising them.
  - The smoke detector senses the drop in current and sets off the alarm.

How a smoke alarm works.

Dating

- Some **rock** types such as granite contain traces of **uranium**, a radioactive material.
  - The **uranium** isotopes present in the rocks go through a series of decays, eventually forming a stable isotope of **lead**.
  - By comparing the amounts of uranium and lead present in a rock sample, its approximate age can be found.

- **Carbon-14** is a radioactive isotope of carbon that is present in all living things.
  - By measuring the amount of carbon-14 present in an archaeological find, its approximate age can be found.

Radiocarbon-dating has shown that the Turin shroud, thought to date from the time of Christ, is only about 500 years old.

Questions

Grades D-C
1. Why might an oil company use a radioactive tracer?
2. Why is it important that the alpha source in a smoke detector has a long half-life?
3. Why would you expect a new rock to contain a bigger proportion of uranium to lead than an old rock?
4. Why can't carbon dating be used to find the age of an iron tool?

Fission

Nuclear power stations

- A **power station** produces **electricity**.
  It needs an **energy source** such
  as coal, oil, gas or nuclear:
  – to **heat water**
  – that **produces steam**
  – that **turns a turbine**
  – to **generate electricity**.

- A **nuclear** power station uses
  uranium as a fuel instead of burning coal, oil or gas to heat the water.

- Natural uranium consists of two **isotopes**, uranium-235 and uranium-238.
  – The '**enriched uranium**' used as **fuel** in a nuclear power station contains
    a greater proportion of the uranium-235 isotope than occurs naturally.

- **Fission** occurs when a large unstable **nucleus** is split up and there is
  a release of energy in the form of **heat**.
  – The heat is used to boil water to produce steam.
  – The pressure of the steam acting on the turbine blades makes it turn.
  – The rotating turbine turns the generator, producing electricity.

Top Tip! Learn the main stages in the production of electricity.

Chain reaction

- A **chain reaction** can carry on for as long as any uranium fuel remains.
  – This allows large amounts of energy to be produced.
  – In a **nuclear power station** the chain reaction is controlled to produce a steady supply of heat.
  – An **atomic bomb** is a chain reaction that has gone out of control.

Artificial radioactivity

- Materials can be made radioactive by putting them into a **nuclear reactor**. They are used:
  – in hospitals to diagnose and treat patients
  – in industry as tracers to detect leaks.

- **Artificial radioisotopes** can be produced by bombarding atoms with the neutrons present in
  a nuclear reactor.
  – Neutrons are uncharged so they're easily captured by **nuclei**, producing **unstable isotopes**.

Radioactive waste

- Nuclear fission produces **radioactive waste** that has to be handled carefully and
  disposed of safely.
  – **Very low-level waste**, such as that produced by medical applications, is placed
    in sealed plastic bags then buried or burned under strict controls.
  – Other **low-level waste** may be embedded in glass discs and buried in the sea.
  – **High-level waste**, such as spent fuel rods, is re-processed to make more
    radioactive materials.

Questions

Grades G-E
1 What fuel is used in a nuclear power station?

Grades D-C
2 What is meant by 'enriched uranium'?

Grades G-E
3 Where are artificial radioisotopes produced?

Grades D-C
4 Why is radioactive waste dangerous?

P4 Summary

P4 RADIATION FOR LIFE

Electrostatics

There are two kinds of electric charge, positive and negative:
- **like charges repel**
- **unlike charges attract**.

Electrostatic effects are caused by the **transfer of electrons.**

You can get an electric shock if you become charged and then become earthed.

Some uses of electrostatics are:
- defibrillators
- paint sprayers
- photocopiers.
- dust precipitators.

Using electricity safely

An electric current is a flow of electric charge. The **potential difference** (pd) between two points in a circuit is the difference in voltage between the two points.

$$\text{resistance } (\Omega) = \frac{\text{pd (V)}}{\text{current (A)}}$$

In a plug:
L = live (brown)
N = neutral (blue)
E = earth (green and yellow stripe)
The **fuse** is connected in the live side. It melts if the current exceeds its stated value, breaking the circuit.

Electric circuits need a complete loop to work.
Resistors are used to change the current in a circuit.

Ultrasound

Medical uses of ultrasound include:
- scans that allow a doctor to see inside you without surgery
- measuring the rate of blood flow in the body
- breaking up kidney or gall stones.

Ultrasound is sound of a higher frequency than we can hear (above 20 000 Hz).

Sound is a longitudinal wave; the particles vibrate in the same direction as the wave travels.
- **Wavelength** (λ) is the distance occupied by one complete wave.
- **Frequency** (f), measured in hertz (Hz), is the number of complete waves in 1 second.
- **Amplitude** is the maximum distance a particle moves from its normal position.

Nuclear radiation

Medical uses of radioactivity include:
- for diagnosis, as a tracer
- to sterilise equipment
- to treat cancers.

Only β- and γ-radiation can pass through skin. γ-radiation is the most widely used for medical purposes as it is the most penetrating. X-rays are similar to γ-rays but are produced in different ways.

Fission is the splitting of a large nucleus, such as uranium, releasing energy. This can set up a chain reaction producing a large amount of energy, as in a nuclear power station or an atomic bomb.

Nuclear radiation is emitted from the nuclei of radioactive materials:
- **α-particle** – helium nucleus
- **β-particle** – fast-moving electron
- **γ-radiation** – electromagnetic waves with very short wavelengths.

Background radiation is ionising radiation that is present in the atmosphere. It's caused by:
- radioactive substances in rocks (especially granite) and soil
- cosmic rays.

A power station uses a source of energy – coal, oil, gas or nuclear fuel:
- to heat water
- to produce steam
- to turn a turbine
- which turns a generator to produce electricity.

A nuclear power station uses uranium as a fuel.

Satellites, gravity and circular motion

Satellites

- A **satellite** is an object that orbits a larger object in space. Satellites stay in **orbit** because of **gravitational attraction**.
- The Moon is a natural satellite. Many other planets also have moons orbiting them.
- Since 1957, artificial satellites have been orbiting Earth and other planets.

Gravity and geostationary satellites

- Every object in the Universe attracts every other object. These forces of attraction are usually only significant when they are on an astronomical scale.
- Planets stay in orbit around the Sun because of gravitational attraction.
- Any object moving in a circle needs a force towards the centre of the circle. This force is called **centripetal** force.
- A **geostationary** satellite orbits above the equator. One orbit takes 24 hours – this means it always appears to be in the same position above the equator.
- Geostationary satellites are used for communications and long-range weather forecasting.
- The nearer a satellite is to the Earth's surface, the shorter its orbit time.

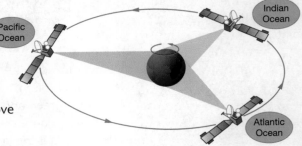

Three geostationary satellites can cover the Earth.

Uses of satellites

- Satellites are used for:
  - **communications**: telephone messages, television programmes and computer data are all transmitted to satellites, where the signals are amplified and re-transmitted back to Earth
  - **weather forecasting**: images from satellites are transmitted back to Earth to improve the accuracy of weather forecasting
  - **GPS**: up to 30 satellites are used to pinpoint a position accurate to 1 m.
- Polar orbit satellites orbit above the North and South poles. Most take about 90 minutes to orbit the Earth.
- Polar orbit satellites can view the whole of the Earth's surface as it rotates beneath them.
- They are used for imaging the Earth – including short-range weather forecasting.

Top Tip! There are many uses of satellites. You will need to explain how they are used.

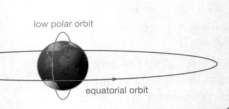

Questions

Grades G-E

1. Ganymede is one of Jupiter's natural satellites. Galileo is an artificial satellite orbiting Jupiter. Which satellite was the first to orbit Jupiter?

Grades D-C

2. Why are geostationary satellites placed in an orbit that takes 24 hours?

Grades G-E

3. Describe one other use of a satellite.

Grades D-C

4. Compare the height of a polar orbit satellite to the height of a geostationary satellite.

Vectors and equations of motion

Relative motion

Top Tip! Remember: always show how you worked out your answer.

- Direction is important if you need to find a particular place.
- If you dropped your wallet or purse within 100 m of a point, it would be easy to find if you knew the direction. If you didn't, you would be looking in an area the size of 7 football pitches.
- On a motorway, a car travelling at 70 mph takes longer to pass a car travelling at 69 mph than it does to pass a car travelling at 30 mph. In the first example, the relative speed is 70 − 69 = 1 mph. In the second example, the relative speed is 70 − 30 = 40 mph.
- If cars are travelling towards each other, the relative speeds are added.

Scalars and vectors

- A **scalar** quantity has magnitude only – a speed of 20 m/s is a scalar.
- A **vector** quantity has both magnitude and direction – a speed of 20 m/s due north is a vector.
- Speed in a particular direction is called **velocity**.
- Vectors are usually represented by arrows in the correct direction; the length of the arrow represents the magnitude of the vector.

    3 m/s to the right ⟶   2 m/s to the left ⟵

- When vectors act in a straight line, they are added algebraically.

    3 m/s to the right ⟶ + 2 m/s to the left ⟵ = 1 m/s to the right ⟶

Measuring speed

- Speed tells us how fast something moves, but not its direction.
- Speed changes during a journey so it is more useful to calculate average speed.

$$\text{speed} = \frac{\text{distance}}{\text{time}}$$

$$\text{average speed} = \frac{\text{total distance travelled}}{\text{total time taken}}$$

Equations of motion

- We use symbols to represent five different quantities:
  - u = initial velocity in m/s
  - v = final velocity in m/s
  - a = acceleration in m/s^2
  - s = distance travelled in m
  - t = time taken in s.

Top Tip! When using equations of motion, make sure you always work in the correct units.

- Final velocity can be calculated using: $v = u + at$
- Distance travelled can be calculated using: $s = \frac{1}{2}(u + v)t$

Questions

Grades G-E

1. Two trains are travelling towards each other. One is travelling at 110 km/h and the other at 130 km/h. What is their relative speed?

Grades D-C

2. Which of these are scalar quantities and which are vector quantities?

    mass   temperature   force   acceleration   volume

Grades G-E

3. Usain Bolt ran a 200 m race in 20 s. Calculate his average speed.

Grades D-C

4. A Renault Espace accelerates from 0 to 30 m/s in 3 s. How far does the car travel in that time?

Projectile motion

Projectiles

- A **projectile** is any object that moves in the Earth's gravitational field.
- The path of a projectile is called its **trajectory**.
- The shape of the curved path is called a **parabola**.

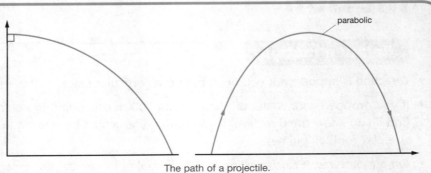

The path of a projectile.

Horizontal projection

- A ball thrown horizontally from the top of a tower will fall to the ground at the same rate as a ball dropped from the top of the tower.
- The ball has a constant horizontal velocity (if you neglect air resistance).
- Both balls accelerate toward the ground.

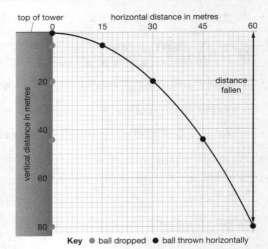

One of Newton's drawings showing his theory of projectiles.

How two balls behave when they are dropped.

- Newton suggested that if the ball could be thrown hard enough from a very tall peak, it would never hit the ground but stay in orbit. Gravity would make the ball fall to Earth but never reach it because of the Earth's curvature.

Analysing projectile motion

- The photograph taken with a stroboscopic light shows that:
  - the horizontal speed is constant
  - the ball slows down as it goes up
  - the ball speeds up as it falls.
- The only force on the ball is a vertical force due to gravity.
- $F = ma$ so there is a downwards acceleration of 10 m/s^2.

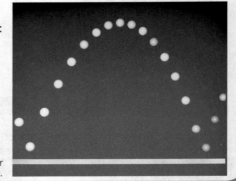

A bouncing ball shown as a stroboscopic image. A light flashing at regular intervals allows successive positions of the ball to be seen.

Questions

Grades G-E

1 Write down one sporting example of a projectile.

Grades D-C

2 When Heather serves at tennis, she strikes the ball very hard in a horizontal direction. What is the horizontal acceleration of the ball as it crosses the court to her opponent?

3 A golf ball has a mass of 45.9 g. Calculate the force acting on it when it is in flight.

Momentum

Force pairs

Top Tip! Remember: every action has an equal and opposite reaction.

- Forces always occur in pairs.
- If one skater pushes the other with a force F, the second skater also pushes with a force F which is the same size but in the opposite direction. The skaters move apart.
- The original pushing force, F, is called the **action**.
- The resulting pushing force, F, is called the **reaction**.

Action and reaction

- An object has **weight** because it is attracted to Earth due to the gravitational attraction of Earth.
- The Earth experiences a reaction force towards the object due to the gravitational attraction of the object.

Momentum

- momentum = mass × velocity
- Momentum is a vector quantity. The units of momentum are kgm/s.

Collisions

- Collisions are not just accidents between cars, trains or aircraft.
- Meteorites are balls of dust and rock that sometimes collide with Earth. Most of them are small and many of them burn up in the atmosphere.
- Scientists think that a large meteorite colliding with Earth millions of years ago led to the extinction of the dinosaurs.
- Scientists sent a spacecraft on a collision course with a comet to find out more about what makes up a comet.

Pool ball and cue – there will be several collisions after the ball is hit.

Avoiding injury

- force = $\dfrac{\text{change in momentum}}{\text{time}}$
- The longer the impact time, the smaller the force.
- Seat belts are designed to increase the length of time it takes for a person to stop if they are in an accident. This means the force on them is less.
- Sports injuries, from jumping, are reduced if the competitor takes longer to stop.

The gymnast bends her knees as she lands.

Questions

Grades G-E
1. Two children are on skateboards. One pushes the other. Why do they both move backwards?

Grades D-C
2. Calculate the momentum of a 0.15 kg cricket ball travelling at 30 m/s.

Grades G-E
3. Suggest three examples of collisions during a football match.

Grades D-C
4. When a parachutist lands, he bends his legs. Explain why.

Satellite communication

Radio waves

- The wavelength of radio waves varies from a fraction of a millimetre up to 10 kilometres.
- High **frequency** radio waves are known as **microwaves**.
- High frequency waves are more penetrating than low frequency waves:
  - high frequency microwaves can pass through the Earth's atmosphere
  - lower frequency radio waves are absorbed or reflected by the Earth's atmosphere.

Communicating with satellites

- Microwave signals are sent into space from a parabolic **transmitter**.
- The signals are received, **amplified** and re-transmitted back to Earth by a geostationary satellite.
- The signals are picked up by a parabolic **receiver**.
- The **ionosphere** reflects radio waves with frequencies below 30 MHz.
- Frequencies above 30 GHz are absorbed and scattered. This reduces signal strength.
- The frequencies used for satellite communication are between 3 and 30 GHz.

Top Tip! Remember: mega means million (10^6), giga means thousand million (10^9).

Diffraction

- Diffraction happens when waves pass through a gap or around a large object.
- An **aerial** detects waves from radio and television transmitters based on Earth.
- A special dish is needed to detect the weak microwave signals from a satellite transmitting television programmes. The waves are then focused on to an electronic device that converts the microwave signals into a useful form.

a Water waves spread out as they pass through a harbour wall.

b Radio waves spread out around a large building.

- The smaller the size of a gap, the greater the diffraction.
- Long and medium wavelength radio waves diffract around buildings, hills and follow the curvature of the Earth.
- TV signals have shorter wavelengths so do not show much diffraction. Television aerials must be in **line of sight** with the transmitter.

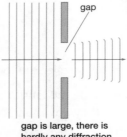
gap is large, there is hardly any diffraction

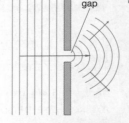
gap is small, waves diffract

Diffraction of waves.

Questions

Grades G-E
1. Why is it that a geostationary satellite cannot receive low-frequency radio waves but a low polar orbiting satellite can?

Grades D-C
2. Why do scientists use frequencies in the range 3–30 GHz to communicate with satellites?

Grades G-E
3. What happens to microwave signals when they reach the surface of the dish?

Grades D-C
4. Most television aerials are on tall poles mounted on chimneys. Explain why.

Nature of waves

Interference

- **Interference** occurs when waves overlap.
- Where waves add together, there is reinforcement. In water waves you would see a big disturbance.
- When sound waves overlap, there are areas which are louder and areas which are quieter.
- When light waves overlap, there are areas which are brighter and areas which are darker.
- If two loudspeakers are connected to the same output of an oscillator and placed about 1 m apart, you will hear alternate loud and quiet areas as you move along a line in front of them.
- When two **crests** or two **troughs** overlap, the waves are in step and the output is a maximum.
- When a trough and a crest overlap, the waves are out of step and the output is a minimum.

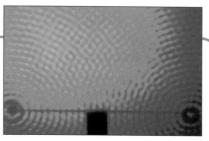
Interference pattern in water waves in a ripple tank.

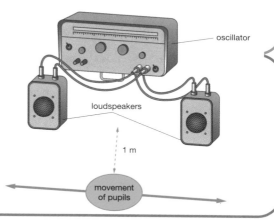

Light travels in straight lines

- Lasers are a source of very concentrated light waves.
- You can see laser light travelling in a straight line.

Light as a wave

- Interference between light waves is produced when light diffracts as it passes through narrow slits.
- The wavelength of light is between 0.0004 and 0.0006 mm so the slits have to be very narrow.
- Interference of light can only be explained if light is a wave.
- Electromagnetic waves, such as light, are transverse waves. Oscillations occur in all directions at right angles to the wave direction.
- Light is **polarised** if the oscillations only occur in one direction at right angles to the wave direction.

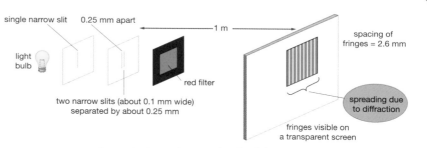
Apparatus to produce overlapping light waves.

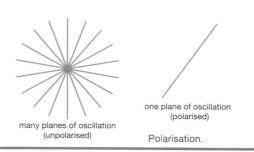
Polarisation.

Questions

Grades G-E
1. What happens when two stones are dropped into a pond?

Grades D-C
2. David is standing between two loudspeakers and can hear a quieter noise than if he moves slightly in either direction. What is happening to the sound waves?

Grades G-E
3. Describe one other piece of evidence that shows light travels in straight lines.

Grades D-C
4. Copy and finish the diagram to show what would happen to light as it passed through a slit if light were a particle and not a wave.

Refraction of waves

Refraction

- **Refraction** occurs when waves pass from one **medium** to another. Refraction often involves a change in the wave direction.
- When light passes from air into glass, the angle of **incidence** is greater than the angle of refraction.
- Different colours of light are refracted by different amounts. Blue light is **deviated** more than red. Sunlight is a mixture of different colours. When sunlight is refracted through raindrops, the colours spread out to form a rainbow – this is **dispersion**.
- Refraction occurs because when light enters a different medium, its speed changes:
  - light slows down as it enters a more dense medium; it deviates towards the normal
  - light speeds up as it enters a less dense medium; it deviates away from the normal.
- The **refractive index** indicates the amount of deviation – the greater the deviation, the higher the refractive index.
- Dispersion occurs because each colour slows down by a different amount when white light enters a medium such as glass, plastic or water, and speeds up by a different amount on leaving.

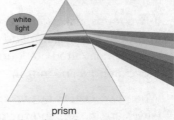

Total internal reflection and critical angle

- When light passes from a more dense medium into a less dense medium:
  - most of the light is refracted but there is a weak reflected ray
  - if the angle of incidence is big enough, all the light is reflected inside the more dense medium.
- When light passes from a more dense medium into a less dense medium the angle of refraction is greater than the angle of incidence.
- The **critical angle**, c, is the angle of incidence in the more dense medium that produces an angle of refraction of 90° in the less dense medium.
- If the critical angle is exceeded, the light is totally internally reflected.
- **Optical fibres** rely on total internal reflection transmitting light along a thin fibre. They are used to carry telephone conversations and computer data as pulses of laser light, and in endoscopes to look inside the body without surgery.

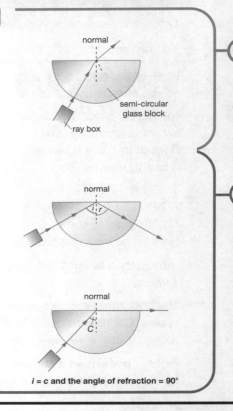

$i = c$ and the angle of refraction = 90°

Questions

Grades G-E

1 Which colour of light is deviated the least?

Grades D-C

2 Light is travelling from glass into air along the normal. Its direction does not change. What happens to its speed?

Grades G-E

3 Light is travelling from air into water. Why is the light not totally reflected?

Grades D-C

4 The critical angle for glass is 41°. Finish the diagram to show the path of a ray of light through an isosceles right-angled prism.

Optics

Convex lenses

- A **convex** or **converging** lens is thinner at its edges than it is in the centre.
- When light strikes a convex lens, it is refracted and brought to a **focus**.
- The focal length is the distance from the centre of the lens to the focus (**focal point**).
- The thicker the lens, the shorter the focal length. Thicker lenses are more powerful.
- Projectors and cameras use convex lenses to produce real images. A **real** image can be projected onto a screen.

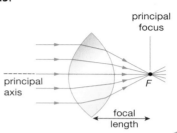

Focal points

- A parallel beam of light from a distant object can be converged to a focus in the **focal plane**.
- If the beam is parallel to the principal axis, the light is focused to the focal point on the principal axis.
- A diverging beam from a near object will focus at a point beyond the focal plane.
- A camera is box with a lens at one end and a film at the other:
  – light from an object passes through the lens and is focused on the film
  – the size of the image is smaller than the object.
- In a projector, the film is placed closer to the lens than the object is to the camera:
  – light from a bulb passes through the film and the lens and is focused on the screen
  – the size of the image is larger than the object.

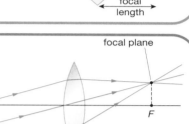

Using convex lenses

- When an object is looked at through a convex lens placed close to the object, the object appears larger. This is a magnifying glass.
- In a digital camera, light is focused onto a light sensitive chip. The information is stored as millions of minute coloured dots called **pixels**.
- A magnifying glass image is the right way up and cannot be projected onto a screen.
- A camera forms a real, inverted, small image on the film. Simple lens cameras have a fixed lens. This only produces a sharp image for one object distance. Better quality cameras have an adjustable lens distance, so the image is sharp whatever the object distance.
- The shutter opens and closes to allow light on to the film. The **aperture** is an adjustable hole allowing different amounts of light onto the film.
- Projectors form a large, inverted, real image on a screen:
  – the lens and/or screen can be moved to produce a sharp image
  – the condenser lenses make sure the film is uniformly illuminated
  – the curved mirror reflects light back to the condenser lenses.

Questions

Grades G-E
1 Projectors, cameras and magnifying glasses use convex lenses. What else uses a convex lens?

Grades D-C
2 Light is brought to a focus at the focal point on the principal axis of a convex lens. Describe the incident light in as much detail as possible.

Grades G-E
3 Jane's camera phone has 8 megapixels. Why is her phone better than one with only 6 megapixels?

Grades D-C
4 What is the advantage of having a camera lens which is adjustable?

P5 Summary

P5 SPACE FOR REFLECTION

Satellites and gravity

Geostationary satellites take exactly 24 hours to **orbit** Earth, above the equator. This means they always stay above the same point on Earth.

Some **radio waves** are reflected by the ionosphere. High **frequency** radio waves (**microwaves**) pass through it to reach an orbiting satellite. Long wavelength radio waves are easily **diffracted** around hills; short wavelength microwaves only diffract a small amount.

Any object moving in a circle has a **centripetal force** that maintains its circular path.

A **satellite** is an object that orbits a larger object in space. A **gravitational force** keeps a satellite in orbit.

Motion

Momentum is a vector.
Momentum = mass × velocity

A **scalar** quantity (e.g. speed) has size only. A **vector** quantity (e.g. velocity) has size and direction.

Equations of motion (for uniform acceleration):
$v = u + at$
$s = \frac{1}{2}(u + v)t$

Projectiles:
– have a constant horizontal velocity
– accelerate towards the ground at 10 m/s^2.

Light and waves

Light travels in straight lines. **Diffraction** and **interference** of light can only be explained by a wave model.
Polarised light has oscillations in one plane only.

Interference occurs when two waves overlap. They can reinforce or cancel. This results in louder and quieter areas in sound and bright and dark areas in light.

Light sometimes produces a rainbow, or spetrum of colours – **dispersion**.

Refraction of waves occurs when light passes from one medium to another. As the speed of the light changes, its direction can change.

The **refractive index** tells us about the amount of bening.

Total internal reflection can occur when light goes from a dense to a less dense medium. The **critical angle** is the angle of incidence for which the angle of refraction is 90°. At greater angles of incidence all the light is totally internally reflected. This happens in an **optical fibre**.

A **convex lens** makes a beam of light converge. Convex lenses are used in cameras and projectors to produce real, inverted images, and as a magnifying glass to produce an image that is the right way up.

Resisting

Basic electric current

- Dimmer switches allow you to control the brightness of a bulb. Some dimmer switches contain a variable resistor.
- The electrical unit used for:
  - voltage is the **volt** (V)
  - current is the **amp** (A)
  - resistance is the **ohm** (Ω).

Top Tip! Always give the units in your answer.

Circuit symbol for a variable resistor.

Resistance and current

- The higher the resistance, the lower the current.
- The higher the current, the brighter the bulb in a circuit.
- The higher the current, the faster a motor turns.
- resistance = voltage ÷ current

Top Tip! Examiners often ask you to draw or label circuit diagrams. It is important to know where ammeters and voltmeters are placed and what they measure.

Electric circuits

- Ammeters are placed in **series** and measure current.
- Voltmeters are placed in **parallel** and measure voltage.
- When the current increases, the voltage across the resistor increases as well.
- When a current passes through a wire, the wire gets hot.
- When a wire gets hot, its resistance increases.

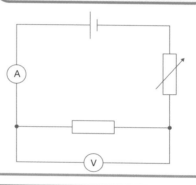

Ohmic and non-ohmic devices

- The voltage–current graph for a resistor is a straight line passing through the origin. This shows that voltage is proportional to current. A resistor obeys Ohm's law.
- The **gradient** of the voltage–current graph is the resistance.
- The voltage–current graph for a bulb is not a straight line passing through the origin. The gradient of the line increases as the current increases. A bulb does not obey Ohm's law.

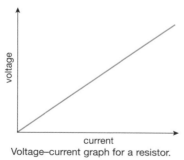

Voltage–current graph for a resistor.

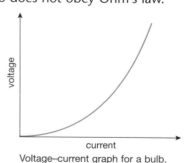

Voltage–current graph for a bulb.

Questions

Grades G-E
1. Write down one use of a variable resistor.

Grades D-C
2. What happens to the brightness of a bulb if the resistance in the circuit is increased?

Grades G-E
3. A fuse wire breaks when the current is too large. What happens to the wire to make it break?

Grades D-C
4. Draw a circuit diagram to show how to measure the voltage–current characteristics for a bulb.

Sharing

Potential divider

- The **voltage** across a **resistor** is sometimes known as **potential difference**.
- The **resistance** of a wire depends on its length – the longer the wire, the greater the resistance.
- The diagram shows how a 10 V potential difference varies across the length of a resistor.
- Some **potential dividers** work by a contact moving over a circular coil – for example, the volume control of a hi-fi unit.

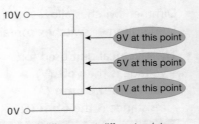

Potential difference at different points along a resistor.

Potential divider circuits

- Some potential divider circuits have fixed resistors.
- The output voltage is a fixed proportion of the supply voltage.
- Some potential dividers have one variable resistor.
- The output voltage can be altered.

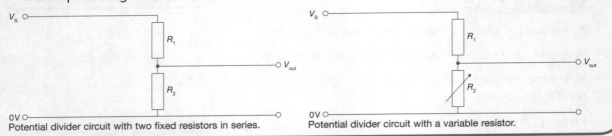

Potential divider circuit with two fixed resistors in series.   Potential divider circuit with a variable resistor.

LDRs and thermistors

- The resistance of a **light-dependent resistor** (LDR) changes as the amount of light shining on it changes.
- Many street lights have LDRs on top of them.
- The resistance of a **thermistor** (thermal resistor) changes when the temperature of its surroundings changes.

- The resistance of an LDR decreases as the light intensity increases.
- The resistance of a thermistor decreases as the temperature increases.

Top Tip!

The behaviour of a thermistor is opposite to that of a resistor. The resistance of a resistor increases as temperature increases.

Graph to show resistance against light in a light-dependent resistor.

Graph to show resistance against temperature in a typical thermistor.

Questions

Grades G-E

1. Write down one use of a potential divider.

Grades D-C

2. A potential divider circuit is made from two identical resistors. A supply voltage of 12 V is connected across them. What are the three possible output voltages?

Grades G-E

3. What property of a thermistor changes when the temperature changes?

Grades D-C

4. The resistance of a LDR on the top of a streetlight is 10 000 000 Ω. Explain why you know it is night time.

Motoring

Field due to a current

- When there is an electric current in a wire, the wire is surrounded by a **magnetic field**.
- The magnetic field can be shown with a plotting compass.
- The magnetic field due to the current in a wire is a circular field around the wire.
- If a wire is in a magnetic field and the current is switched on, the two magnetic fields interact and the wire moves.

Field direction

- The direction of the magnetic field around a wire can be found by using the **right-hand grip rule**.
- The field pattern due to a long coil of wire is similar to that of a bar magnet.
- When a wire is placed between the poles of a magnet, the wire moves out of the gap when the current is switched on.
- If the direction of the current is reversed, the wire moves in the opposite direction.

Imagine gripping the wire with your right hand, with your thumb pointing in the direction of the current. Your fingers point around the wire in the direction of the field

The right-hand grip rule used to predict a magnetic field around a wire carrying a current.

North pole — end view — anticlockwise

South pole — clockwise

Top Tip!
You can work out the North and South ends of the coil by seeing which way the current passes.

Uses of motors

- Many electrical appliances have **motors** in them. Motors make things turn.

Top Tip!
Examiners may ask you to be specific about the use of a motor, e.g. the turntable in a microwave oven, not simply a microwave oven.

Turning coils

- When a current passes through a coil, placed between the poles of a magnet, there is a force on each side of the coil.
- Using **Fleming's left-hand rule**, you can see that the force on each side of the coil is in the opposite direction. As one side is forced up, the other side is forced down.
- The coil starts to spin.
- The motor spins faster when:
  - the number of turns on the coil is increased
  - the size of the current is increased
  - the strength of the magnetic field is increased.

A simple electric motor.

Questions

Grades G-E
1 Draw a diagram to show the magnetic field due to a wire carrying an electric current.

Grades D-C
2 Draw a diagram to show the magnetic field due to a long coil of wire carrying an electric current.

Grades G-E
3 Describe how an electric motor is used in one common kitchen appliance.

Grades D-C
4 What is the job of the brushes in an electric motor?

Generating

The dynamo effect

- Moving a magnet near to a wire or moving a wire near to a magnet produces a current in the wire.
- When a coil is rotated between the poles of a magnet, a current is produced in the coil.
- The direct current **dynamo** is a motor working backwards.
- Mains electricity is produced using alternating current **generators**.
- The **frequency** of the alternating current supply is 50 Hz.

Making a current in a wire.

Induced voltages

- When a wire is held stationary between the poles of a magnet, there is no current in the wire.
- If the wire is moved upwards, a voltage is **induced** in the wire and a current passes. The same thing happens if the magnet is moved downwards. It is the relative movement of wire and field that is important.
- When the wire moves downwards or the magnet moves up, the induced voltage is reversed and the current is in the opposite direction.
- Reversing the direction of the magnetic field also reverses the induced voltage and current direction.
- A voltage is induced whenever a magnetic field changes.

Alternating current generators

- The bicycle dynamo contains a permanent magnet rotating inside a coil.
- As the magnetic field through the coil changes, a changing current (AC) is produced in the coil.
- In a power station generator, the magnetic field inside the coil is produced by electromagnets.
- The electromagnet is made from coils of wire (**rotor** coils) which are turned by the **turbine**.
- An alternating current is produced in the **stator** coils which surround the rotor coils.

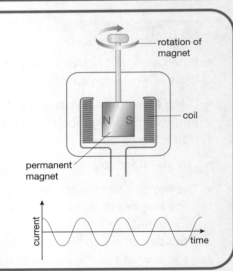

A dynamo and the alternating current it generates.

Questions

Grades G-E

1 David has built a model motor. He spins the coil. What effect does this have?

Grades D-C

2 Shelly connects the ends of a piece of wire to an ammeter. She moves the wire downwards between the poles of a magnet. The ammeter needle moves to the left. She reverses the direction of the magnetic field and moves the wire upwards between the poles. In which direction does the ammeter needle move?

3 What happens to the current in the stator coils and the frequency of the current if the turbine turns the rotor coils faster?

Transforming

Voltage changer

Top Tip! Transformers do not change alternating current into direct current.

- A **transformer** changes the size of an alternating current voltage.
- A **step-down** transformer decreases the voltage.
- You would use a step-down transformer in a mobile phone charger, radio and laptop.
- A **step-up** transformer increases the voltage.
- An **isolating** transformer does not change the voltage.
- Isolating transformers are used in shaver sockets.

Transformer design

- A transformer consists of two coils of wire wound onto an iron **core**.
- The input alternating current voltage is connected to the **primary** coil.
- The output alternating current voltage is obtained from the **secondary** coil.

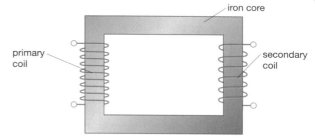

A simple transformer.

$$\frac{\text{primary voltage}}{\text{secondary voltage}} = \frac{\text{number of turns on primary coil}}{\text{number of turns on secondary coil}}$$

- Step-down transformers have more turns on the primary coil.
- Step-up transformers have more turns on the secondary coil.
- Isolating transformers are used for safety – the water and steam in the bathroom could lead to electrocution or damage to the house wiring system if an ordinary socket is used.

The National Grid

- Power stations generate electricity at 25 000 V.
- Step-up transformers increase the voltage to 400 000 V.
- The **National Grid** is the system that transmits electricity around the country.
- Sub-stations have step-down transformers to reduce the voltage. In the home this is 230 V.

Energy loss

- When a current passes through a wire, the wire gets hot.
- The overhead power lines of the National Grid get hot and lose energy to the surroundings.
- Power is a measure of how fast energy is transferred.
  power loss $\propto$ current^2

Questions

Grades G-E
1. What type of transformer is used in the charger for an MP3 player?

Grades D-C
2. A transformer for a mobile phone has 20 times as many coils on the primary coil as on the secondary coil. The supply voltage is 230 V. What is the output voltage?

Grades G-E
3. What type of transformer changes the 25 000 V from the power station to 400 000 V?

Grades D-C
4. The current in a wire is increased from 1 A to 3 A. By how much is the power loss increased?

Charging

Diodes

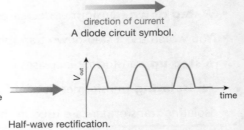
direction of current
A diode circuit symbol.

- A diode is an electronic component that only allows a current to pass in one direction.
- A single diode is used in a circuit to change alternating current into direct current.
- The direct current is not constant but it is always in the same direction. This is **half-wave rectification**.

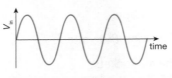
Half-wave rectification.

- Use the same circuit for diode characteristics as you would for a resistor or a bulb.

Top Tip! Remember to reverse the power supply when examining diode characteristics.

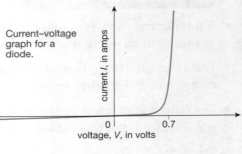
Current–voltage graph for a diode.

- Most diodes need a voltage of about 0.6 V before they start to work.
- When a diode is connected so that a current passes, it is **forward biased**.
- When a diode is connected so that no current passes, it is **reverse biased**.
- A resistor is always used in series with a diode to protect the diode.

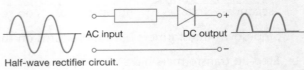
Half-wave rectifier circuit.

- Four diodes can be arranged in a special circuit that will allow a continual direct current to pass. This is **full-wave rectification**.

Full-wave rectification.

Capacitors

Capacitor circuit symbols.

- A **capacitor** is an electronic component that stores electric charge.
- Four diodes can be arranged to make a bridge circuit. The addition of a large capacitor makes the output smoother.

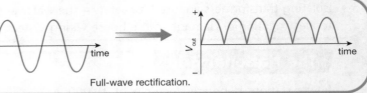

- When a direct current supply is connected to a capacitor, the capacitor becomes charged. The voltage across the capacitor increases until it is equal to the supply voltage.
- When the capacitor is connected to a resistor, for example, the voltage decreases as the capacitor discharges.

Top Tip! The time it takes for a capacitor to discharge depends on the resistance and the capacitor.

Questions

Grades G-E

1. Jane and John each connect a circuit containing a battery, resistor, diode and bulb. The bulb in Jane's circuit lights but the bulb in John's circuit does not. Why does the bulb in John's circuit not light?

Grades D-C

2. Draw a circuit diagram to show how you would obtain the characteristics of a diode.

Grades G-E

3. What is the job of a capacitor?

Grades D-C

4. Sketch graphs to show how the voltage across a capacitor changes as it is
   a being charged
   b discharging through a resistor.

It's logical

Logic signals

- Logic gates need a signal of about 5 V to make them work.
- The signal is described as being high and is often represented by 1.
- The absence of a signal is described as being low and is often represented by 0.

NOT gate

- The **NOT** gate has one input and one output.
- The output is the opposite of the input.
- Truth tables show the way in which logic gates behave.

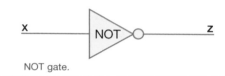

NOT gate.

AND and OR gates

- **AND** and **OR** gates have two inputs.
- The output from an AND gate is high if both inputs are high.
- The output from an OR gate is high if one or both inputs are high.

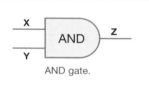

AND gate.

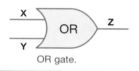

OR gate.

Switching logic gates

- Switches, LDRs and thermistors are used in a potential divider circuit to make the input to a logic gate high.

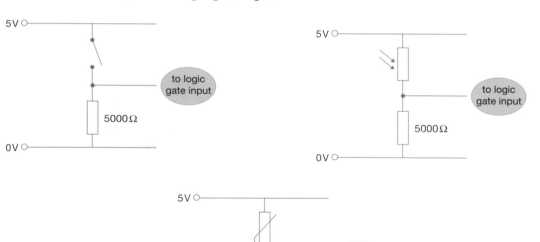

Questions

Grades G-E

1. Write down the truth table for a NOT gate.

Grades D-C

2. Write down the truth tables for **a** an AND gate **b** an OR gate.

3. Look at the potential divider circuit diagram showing a thermistor. What is the resistance of the thermistor when the temperature is cold? What happens to the resistance when the temperature warms up?

Even more logical

Logic circuits

- Logic gates can be combined to make a logic circuit.
- The output from one gate can be used as an input to another.
- The combination is known as an electronic **system**.

Truth tables

- When there are several logic gates combined together, truth tables can be used to work out what happens in the system.

Top Tip!
When you work out a truth table for a system, do it one step at a time. Write out all of the possible input values and then work out what happens at each logic gate.

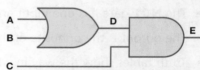

In this example, **A** and **B** are the input signals to the OR gate. **D** is the output signal from the OR gate but is an input signal to the AND gate and so is **C**. The output signal from the AND gate is **E**.

LEDs

- **Light-emitting diodes** (LEDs) are often used as indicator lights.
- LEDs are increasingly being used instead of bulbs.
- The output current from a logic gate can light a LED but not a bulb.

The circuit symbol for an LED.

Relays

- The output current from a logic gate can operate a **relay**. The relay can switch something that needs a bigger current to operate.

- When a current passes through the coil, the iron armature is attracted.
- The armature pivots and pushes an insulating bar against the central contact.
- The central contact moves, opening the normally closed contacts and closing the normally open contacts.

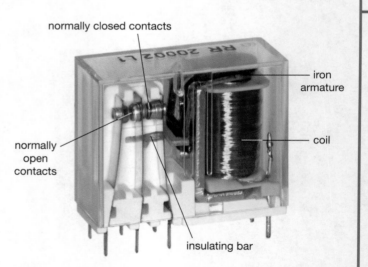

Questions

Grades G-E
1. What is meant by the term 'electronic system'?

Grades D-C
2. Write down the truth table for the electronic system shown that contains an OR gate and an AND gate.

Grades G-E
3. Why must there be a relay between the output from a logic circuit and a light bulb?

Grades D-C
4. Why is the armature attracted towards the coil when a current passes through the coil?

P6 Summary

Electric circuits

- A **potential divider** consists of two resistors used to prouce a specific voltage.
- **Current** is measured in amps (A). **Voltage** is measured in volts (V)
- A **diode** only allows a current to pass in one direction.
- The resistance of a **light-dependent resistor** (LDR) decreases as the light intensity increases.
- The resistance of a **thermistor** decreases as temperature increases.
- Diodes can be arranged as a **rectifier** to change alternating current into direct current.
- **Resistors** control the current in a circuit. resistance = voltage ÷ current. The resistance of a wire increases as the wire gets hotter. Resistance is measured in ohms (Ω).
- A **capacitor** stores charge.

Electromagnetism

- A current-carrying wire has a magnetic field around it. A **motor** moves because of the interaction between the magnetic field due to a coil and a permanent magnet.
- **Generator** A current is induced in a wire when there is a changing magnetic field.
- A **transformer** changes the size of an AC voltage. Electricity is transmitted at high voltages to reduce the energy loss in the overhead power lines.

Logic circuits

- The behaviour of NOT, AND and OR **logic gates** can be described using **truth tables**. Inputs and outputs are described as high (1) and low (0).
- A **relay** uses a small current to switch a larger current. The output from a logic gate is often to a relay which switches a larger current.
- Logic gates work at low voltages (about 5–6 volts). Inputs to logic gates can be controlled by potential divider circuits.

P6 ELECTRICITY FOR GADGETS

How science works

Understanding the scientific process

As part of your Physics assessment, you will need to show that you have an understanding of the scientific process – HOW SCIENCE WORKS.

This involves examining how scientific data is collected and analysed. You will need to evaluate the data by providing evidence to test ideas and develop theories. Some explanations are developed using scientific theories, models and ideas. You should be aware that there are some questions science cannot answer.

Collecting and evaluating data

You should be able to devise a plan that will answer a scientific question or solve a scientific problem. In doing so, you will need to collect data from both primary and secondary sources. Primary data will come from your own findings – often from an experimental investigation. Whilst working with primary data, you will need to show that you can work safely and accurately, not only on your own but also with others.

Secondary data is found by research – often using ICT but do not forget books, journals, magazines and newspapers are also sources of secondary data. The data you collect will need to be evaluated for its validity and reliability.

Presenting information

You should be able to present your information in an appropriate, scientific manner. This means being able to develop an argument and come to a conclusion based on recall and analysis of scientific information. It is important to use both quantitative and qualitative arguments.

Changing ideas and explanations

Many of today's scientific and technological developments have both benefits and risks. The decisions that scientists make will almost certainly raise ethical, environmental, social or economic questions. Scientific ideas and explanations change as time passes and it is the job of scientists to validate these changing ideas.

In 1692, the British astronomer Edmund Halley (after whom Halley's Comet was named) suggested that the Earth consisted of four concentric spheres. He was trying to explain the magnetic field that surrounds the Earth. There was, he said, a shell about 500 miles thick, two inner concentric shells and an inner core. The shells were separated by atmospheres and each shell had magnetic poles. The spheres rotated at different speeds. He believed this explained why unusual compass readings occurred. He also believed that each of these inner spheres supported life which was constantly lit by a luminous atmosphere.

This may sound quite an absurd idea today, but it is the work of scientists for the past 300 years that has developed different models that are constantly being refined.

How science works

Science in the News

Assessment

Science in the News is intended as the main way in which the OCR Physics course assesses your understanding of HOW SCIENCE WORKS.

Whilst some of you will continue to study science, many of you will have completed your science education by the time you have finished your GCSE course. It is important that you are able to meet any scientific challenge which arises in later life.

It is important that you realise when data or information is not presented in an accurate way. Think about what is wrong in this example from a shopping catalogue.

> **Solar Airship**
>
> From box to air in under 2 minutes! Simply unroll the airship and watch it magically inflate as the black surface attracts heat from even minimal sunshine. Tie off one end with the cord provided and fly like a giant 8 metre sausage-shaped kite! Good for all year round use. Folds away into box provided. Comes with a 30 m tether line. Suitable for ages 12+.

You should also be aware of what aspects of science may be important for people living in the 21st Century.

One of the most controversial topics at this time is the use of wind farms for energy production.

Your Science in the News assessment will ask you to undertake some research on a scientific issue. The task set to you will be in the form of a question. You will then have to produce a short report which will clearly show that you have

- considered both sides of the argument
- decided on the suitability, accuracy and/or reliability of the evidence
- considered the impact on society and the environment
- justified your conclusion about the question asked.

The aim is to equip you with life-long skills that will allow you to take a full and active part in the scientific 21st Century.

Collins

Collins Revision

GCSE Foundation Physics

Exam Practice Workbook

FOR OCR GATEWAY B

Heating houses

1 a Finish the sentences by choosing the **best** words from this list.

degrees Celsius   energy   joules   kilograms   power   watts

Temperature is a measure of hotness and is measured in _____ .

Heat is a form of _____ and is measured in

_____ . [3 marks]

b Kelly opens the front door on a very cold morning. Her mother complains that the house is getting cold. Use your ideas about energy flow to explain why the house gets cold.

_____

_____ [2 marks]

2 a A block of copper of mass 100 g is heated in a Bunsen flame. It takes 2 minutes to increase the temperature of the copper by 30 °C. How long will it take to increase the temperature of a 200 g block of copper by 60 °C? Put a ring around the correct answer.

30 seconds   **1 minute**   **2 minutes**   **4 minutes**   **8 minutes**   [1 mark]

b A block of iron of mass 100 g is heated in the same Bunsen flame. Why does it **not** take 2 minutes to increase the temperature of the iron by 30 °C?

_____ [1 mark]

c Finish the sentence.
The energy needed to raise the temperature of 1 kg of a material by 1 °C is known as the _____ . [1 mark]

3 a What happens to a lead block at its melting point?

_____ [1 mark]

b What happens to water at its boiling point?

_____ [1 mark]

c Write down **one** example where energy is transferred but there is no change in temperature.

_____ [1 mark]

d What physical quantity is measured in units of J/kg?

_____ [1 mark]

Keeping homes warm

1 Dan wants to reduce the heating bills for his old house. He decides to insulate his loft and then replace his windows with double glazing.

a Loft insulation contains trapped air. Why does loft insulation reduce energy loss from the house?

_____ [1 mark]

b Write down **three** other ways Dan could insulate his house and reduce energy loss.

_____
_____
_____
_____ [3 marks]

c Dan spends £120 on loft insulation. He is told that this will reduce his heating bills by £40 per year. Calculate the payback time for loft insulation.
Show how you work out your answer.

_____ [2 marks]

d Why does Dan decide to insulate the loft before replacing his windows?

_____
_____ [1 mark]

e Dan heats his house with coal fires. He is told that his fires are **32% efficient**. Explain what is meant by 32% efficient.

_____
_____ [2 marks]

f Why are coal fires so inefficient?

_____
_____ [1 mark]

How insulation works

1 On a cold winter's day, Marc wears a thick waterproof coat and Daniel wears a thin hoody. Marc feels cold but Daniel stays warm. Finish the sentences to explain why Daniel stays warm.

The hoody contains _____ air. Air is a good _____ .

[2 marks]

2 The diagram represents the movement of air around a room. Show where the room heater is by writing the letter **F** on the diagram. [1 mark]

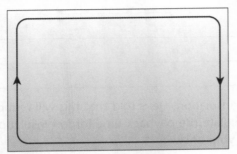

3 a The diagram shows a section through a double-glazed window. Michael says that it is just as effective to use a piece of glass twice the thickness. Use your ideas about energy transfer to explain why double glazing is better.

space filled with air or argon, or has a vacuum

_____
_____
_____
_____
_____

[3 marks]

b New homes are built with insulation blocks in the cavity between the inner and outer walls. The blocks have shiny foil on both sides.

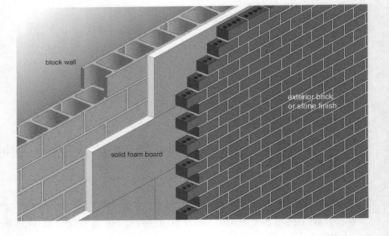

block wall

exterior brick or stone finish

solid foam board

  i Explain how the insulation blocks reduce energy transfer by conduction and convection.

_____
_____
_____
_____

[4 marks]

  ii Explain how the shiny foil helps to keep a home warmer in winter and cooler in summer.

_____
_____

[2 marks]

Cooking with waves

1 a Finish the sentences by choosing the **best** words from this list.

absorb	aluminium	induction	infrared
reflect	spectrum	ultraviolet	water

Warm and hot bodies emit _____ radiation.

Dark surfaces _____ more radiation than light surfaces.

Microwaves are part of the electromagnetic _____ .

_____ molecules absorb microwaves.

[4 marks]

b Why do microwave ovens take less time to cook food than normal ovens?

_____ [1 mark]

2 a Young people are advised not to use mobiles phones too much. Texting is preferable to using them to make phone calls. Why is this advice given to young people?

_____
_____ [2 marks]

b Microwaves are suitable to communicate with spacecraft thousands of kilometres away while sometimes a mobile phone cannot receive a signal just a few kilometres from the nearest transmitter. Why do microwave signals seem to work better in space than they do on Earth?

_____
_____ [2 marks]

c Dave says that microwave signals 'bounce' off satellites. Jackie says Dave is wrong. What happens to a microwave signal when it is received by a satellite?

_____
_____ [2 marks]

Infrared signals

1 a Answer **true** or **false** to each of these statements.

Infrared radiation is part of the electromagnetic spectrum. _____

Passive infrared sensors emit infrared radiation to detect burglars. _____

A remote controller for a CD player emits infrared radiation. _____

[3 marks]

b The diagram shows a signal displayed on an oscilloscope. What type of signal is it? Put a (ring) around the correct answer.

    background   digital   on/off   radial   square   [1 mark]

c Draw a diagram to represent an **analogue** signal.

[1 mark]

2 a A ray of laser light is shone into one end of an optical fibre.

Finish the path of the ray as it passes into, through and out of the optical fibre. [2 marks]

b The diagrams show three rays of light travelling from water into air. The three angles of incidence are (i) smaller than the critical angle (ii) equal to the critical angle (iii) larger than the critical angle.

(i)   (ii)   (iii)

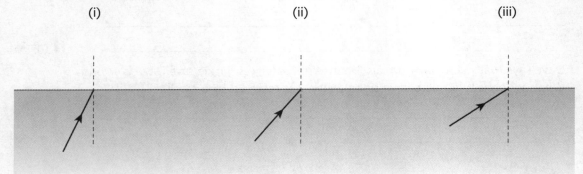

  i Finish the diagrams to show what happens to the rays of light as they continue towards the water/air and after they meet the water/air boundary. [5 marks]

  ii Label the critical angle on the correct diagram with the letter **c**. [1 mark]

Wireless signals

1 a Write down **one** advantage a mobile phone has over a land-line house phone.

_____

[1 mark]

b When Ruby is watching television, she notices that there is a feint second picture slightly offset to the main picture.

Finish the sentence to explain why there is this 'ghost' picture.
Choose the **best** word from this list.

      absorbed   **dispersed**   **reflected**   **refracted**

The aerial has received a direct signal from the transmitter and a signal that has been _____ .   [1 mark]

c Ruby listens to her favourite radio station. Every so often, she notices that she can hear a foreign radio station as well.
Put ticks (✓) in the boxes next to the **two** statements that explain why this happens.

The foreign radio station is broadcasting on the same frequency.	
The foreign radio station is broadcasting with a more powerful transmitter.	
The radio waves travel further because of weather conditions.	
Ruby's radio needs new batteries.	

[2 marks]

d Finish the sentences by choosing the **best** words from this list.

      amplitude   **atmosphere**   **frequency**   **wavelength**

Radio waves are refracted in the upper _____ .

There is less refraction if the _____ is higher.

[2 marks]

Light

1 a Light is a transverse wave that travels at a speed of 300 000 km/s.
Write down the names of **four other** types of transverse wave that travel at 300 000 km/s.

_____
_____
_____
_____ [4 marks]

b The diagram represents a water wave. Water is a transverse wave.
The arrow shows the direction of motion of the **wave**.

Add another arrow to the diagram to show the direction **water particles** move. [1 mark]

2 The diagram shows a transverse wave.

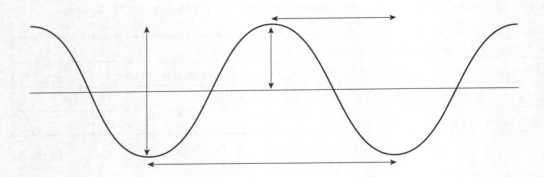

a Write the letter **A** next to the arrow which shows the amplitude of the wave. [1 mark]

b Write the letter **W** next to the arrow which shows the wavelength of the wave. [1 mark]

c What is meant by the **frequency** of a wave?

_____
_____ [1 mark]

3 a Write down **one** way of sending a message over a long distance without using electricity or radio waves.

_____ [1 mark]

b Why was it necessary for Samuel Morse to devise a code of dots and dashes?

_____ [1 mark]

Stable Earth

1 a The diagram shows a device used to detect and measure the strength of earthquakes.

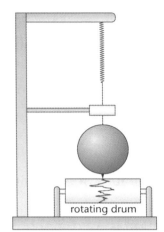

rotating drum

What is this device called? Put a (ring) around the correct answer.

joulemeter   **newtonmeter**   **seismometer**   **wattmeter**   [1 mark]

b Finish the sentences by choosing the **best** words from this list.

fault   **plate**   **shock**   **ultraviolet**   **water**

Earthquakes happen at a _____.

_____ waves travel through and round the Earth. [2 marks]

c **P** waves and **S** waves are two of the waves which travel through the Earth after an earthquake. Put a tick (✓) in the box or boxes to correctly describe each wave. The first one has been done for you.

description	P wave	S wave
pressure wave	✓	
transverse wave		
longitudinal wave		
travels through solid		
travels through liquid		

[5 marks]

2 a What effect do greenhouse gases have on the Earth?

_____ [1 mark]

b When a fuel burns, a greenhouse gas is released. What is the name of this gas?

_____ [1 mark]

c Why does dust from factory chimneys cause warming of the Earth?

_____ [1 mark]

d Leah wants to sunbathe and get a good tan.

  i Why must she be careful not to stay in the Sun for too long?

_____ [1 mark]

  ii She uses a sun screen with **SPF 30**. What does SPF 30 mean?

_____
_____ [2 marks]

P1 Revision checklist

P1 ENERGY FOR THE HOME

- I know the difference between temperature and heat. ☐

- I can explain what is meant by specific heat capacity and specific latent heat. ☐

- I can describe different forms of domestic insulation and explain how they work. ☐

- I can calculate energy efficiency. ☐

- I know the parts of the electromagnetic spectrum and their properties. ☐

- I can describe how infrared radiation is used for cooking and for remote control devices. ☐

- I can describe how microwaves are used for cooking and for communication. ☐

- I know the difference between analogue and digital signals. ☐

- I can describe total internal reflection and its use in optical fibres. ☐

- I can describe the use of wireless signals for communication. ☐

- I know why there is sometimes interference with radio signals. ☐

- I know the properties of a transverse wave and that light is an example of a transverse wave. ☐

- I can describe three types of earthquake wave and how they are detected. ☐

- I know some of the effects of natural events and human activity on weather patterns. ☐

Collecting energy from the Sun

1 The Sun is the source of energy for the Earth.
Write down **two** ways in which the Sun provides plants with the energy they need.

_____
_____ [2 marks]

2 a Finish the sentence by choosing the **best** word from this list.

        chemicals   **electricity**   **heat**   **light**

A photocell uses _____ to produce electricity. [1 mark]

b Write down **four** advantages of using photocells.

_____
_____
_____
_____ [4 marks]

c What happens to the amount of electricity produced if part of the photocell is covered over.

_____ [1 mark]

3 a Solar panels use energy from the Sun to heat water. The solar panel is black.
Water from the pipes in the solar panel is stored in a cylinder.

  i Why is the solar panel black?

_____ [1 mark]

  ii How is energy transferred from the water in the pipes to the storage cylinder?
Put a ring around the correct answer.

      conduction   **convection**   **radiation** [1 mark]

b During the day, energy from the Sun passes through the large window and warms the room; this is called passive solar heating.
How is the room heated during the night?

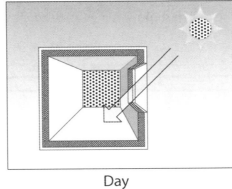

Day

_____
_____ [1 mark]

Generating electricity

1 a The diagram shows a wire connected to an ammeter moving upwards between the poles of a magnet. The needle on the ammeter moves to the right. What happens when the wire is moved **downwards** between the poles of the magnet?

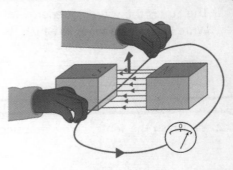

_____

_____

[1 mark]

b The diagram shows a model dynamo. When the coil is spun, a current is produced. Write down **two** ways in which the size of the current can be increased.

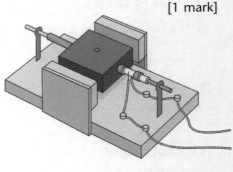

_____

_____

_____

_____

[2 marks]

2 A model generator consists of a coil of wire rotating between the poles of a magnet. How is the structure of a generator at a power station different from the model generator?

_____

[1 mark]

3 The flow diagram represents how fossil fuels at a power station provide electrical energy for distribution around the country.

| fossil fuel burned | → | water heated to produce steam | → | steam turns turbine | → | A | → | generator produces electricity | → | electricity distributed |

a One step in the process has been missed out. What should be in box **A**?

_____

[1 mark]

b How is energy lost from the overhead power lines as electricity is distributed around the country?

_____

[1 mark]

4 a What is the job of a transformer? Put a tick (✓) in the box next to the correct answer.

change the size of an AC voltage	
change the size of a DC voltage	
change AC into DC	

[1 mark]

b High voltage transmission lines distribute electricity around the country at 400 kV. We use electricity in our homes at 230 V. Why is electricity not distributed at 230 V?

_____

[1 mark]

Fuels for power

1 a Power stations use fuels. Some are **renewable** and some are **non-renewable**. Put ticks (✓) in the boxes next to the renewable fuels.

coal	
natural gas	
oil	
straw	
wood	

[2 marks]

b A nuclear power station uses uranium as its energy source. Uranium is not a fuel.
  i What is a fuel?
  _____ [2 marks]
  ii How does uranium provide energy in the form of heat?
  _____ [1 mark]

2 a A nuclear power station has two advantages over a fuelled power station. No smoke is produced and it does not produce carbon dioxide. Write down **one** disadvantage of using a nuclear power station.
_____ [1 mark]

b Finish the sentences by choosing the **best** words from this list.

calcium   cancer   DNA   kinetic   nuclear   mutate   react

Radiation from _____ energy sources causes ionisation.

This causes a change in the structure of atoms. One of the chemicals in

body cells is _____ and when this is exposed to radiation

it can _____. As a result, body cells may divide in an

uncontrolled way. This can lead to _____. [4 marks]

3 a Each of the headlamp bulbs in Anna's car is connected to a 12 V battery. When she switches on the headlamps, a current of 2 A passes through the bulb. Calculate the power rating of the bulb.

_____ [4 marks]

b In her home, Anna uses a 2.5 kW kettle for $\frac{1}{2}$ hour each day. Electricity costs 12p per kWh. How much does it cost Anna to use her kettle each day?

_____ [4 marks]

Nuclear radiations

1 a The picture shows an instrument used to measure radioactivity. What is the name of this instrument?

_____

_____

[1 mark]

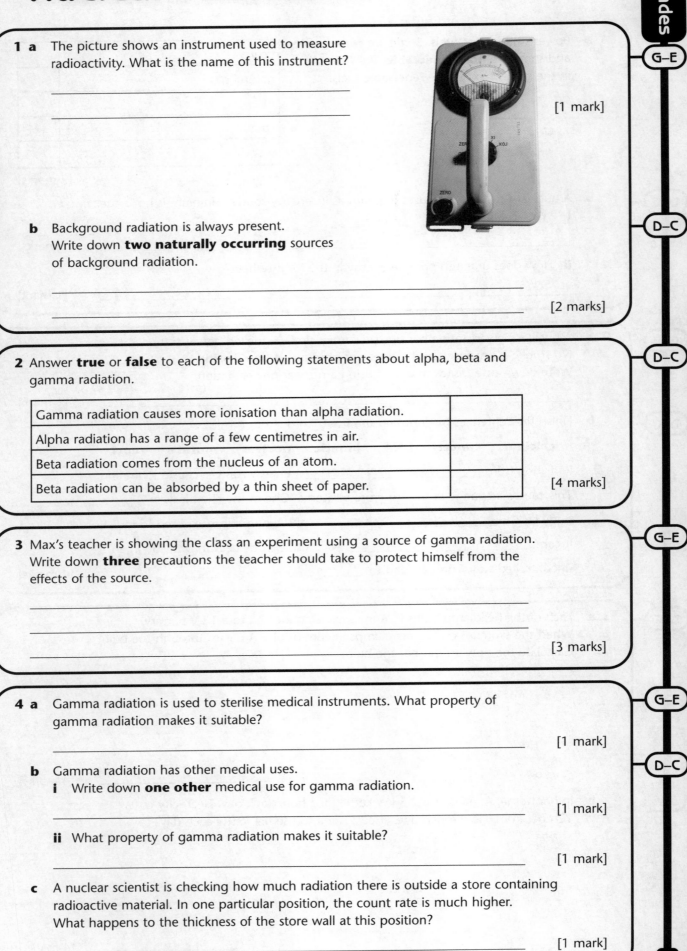

b Background radiation is always present. Write down **two naturally occurring** sources of background radiation.

_____

_____

[2 marks]

2 Answer **true** or **false** to each of the following statements about alpha, beta and gamma radiation.

Gamma radiation causes more ionisation than alpha radiation.	
Alpha radiation has a range of a few centimetres in air.	
Beta radiation comes from the nucleus of an atom.	
Beta radiation can be absorbed by a thin sheet of paper.	

[4 marks]

3 Max's teacher is showing the class an experiment using a source of gamma radiation. Write down **three** precautions the teacher should take to protect himself from the effects of the source.

_____

_____

_____

[3 marks]

4 a Gamma radiation is used to sterilise medical instruments. What property of gamma radiation makes it suitable?

_____

[1 mark]

b Gamma radiation has other medical uses.
  i Write down **one other** medical use for gamma radiation.

_____

[1 mark]

  ii What property of gamma radiation makes it suitable?

_____

[1 mark]

c A nuclear scientist is checking how much radiation there is outside a store containing radioactive material. In one particular position, the count rate is much higher. What happens to the thickness of the store wall at this position?

_____

[1 mark]

Our magnetic field

1 a The magnetic field around Earth is similar to the field around a bar magnet. The North pole of a compass points towards the North magnetic pole. The diagram shows a model of Earth's magnetic field with a bar magnet inside Earth.
Write the letter **N** on the bar magnet to show the position of the North pole of the **magnet**.

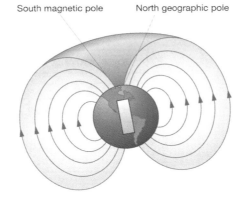

South magnetic pole   North geographic pole

[1 mark]

b Which statement **best** describes the magnetic field due to a coil of wire?
Put a tick (✓) in the box next to the correct answer.

The magnetic field is circular.	
The magnetic field is radial like the spokes of a bicycle wheel.	
The magnetic field is similar to the field due to a bar magnet.	

[1 mark]

2 There are many artificial satellites orbiting Earth. Some provide information about the weather.
Write down **two** other uses of artificial satellites.

_____

_____ [2 marks]

3 a Solar flares erupt from the surface of the Sun. What effect does this have on Earth?
Put a tick (✓) in the box next to the correct answer.

Communication signals are disrupted.	
Earth's average temperature increases by 10 °C.	
The Moon appears much brighter in the night sky.	

[1 mark]

b Finish the sentence by choosing the **best** words from this list.

charged particles   **gamma rays**   **hydrogen atoms**

Solar flares emit _____ that produce magnetic fields.
The magnetic fields interact with Earth's magnetic field.

[1 mark]

4 Scientists believe that the Moon and Earth used to rotate much faster than they do now.
What do they think caused the rotation to slow down?

_____ [1 mark]

Exploring our Solar System

1 a When we look into the sky, there are some things we see because they produce their own light. Some things we see because they reflect light. Some things it is impossible to see at all.

The lists show some of the objects in the sky and a description of how the object may be seen. Draw a **straight** line from each object to the correct way it may be seen. One has been done for you.

object
artificial satellite
black hole
comet
meteor
moon
star

how object may be seen
object cannot be seen at all
object seen because it produces its own light
object seen because it reflects light

[5 marks]

b On 24 August 2006, the International Astronomical Union considered a proposal to redefine planets. There would be twelve planets in our Solar System. Ceres, the largest asteroid would become a planet. Charon, at the moment known as Pluto's moon, would become a 'twin planet' with Pluto. Recently, another planet has been discovered beyond Pluto. This proposal was rejected and the decision made that Pluto should no longer be called a planet.

The diagram shows the Sun, the eight planets, Pluto and the asteroid belt.

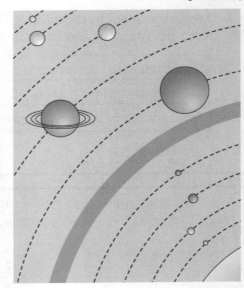

i Write the letter **C** on the diagram to show where Ceres orbits.

ii Write the letter **P** on the diagram to show Pluto.

[2 marks]

2 Scientists are exploring space to find out if there are any other forms of life elsewhere in the Universe. Manned and unmanned spacecraft have been sent into space to investigate.

a How have scientists tried to find out if there are other life forms **without** sending a spacecraft into space?

_____

[1 mark]

b Helen Sharman was the first British astronaut. She spent eight days in the Mir Space Station doing science experiments. When she was in the Space Station she experienced **weightlessness** but was not **weightless**.
Why was Helen never weightless?

_____

[1 mark]

c When astronauts work outside a spacecraft, they have to wear special helmets with sun visors. Why do the helmets need special visors?

_____

[1 mark]

Threats to Earth

1 a An asteroid hit Earth about 65 million years ago making a very large crater.
What else happened when the asteroid hit Earth? Put ticks (✓) in the boxes next to the **three** correct answers.

all life became extinct	
it got colder	
the first human appeared on Earth	
the Moon was formed	
there were a lot of fires	
tsunamis flooded large areas	

[3 marks]

b Most asteroids orbit the Sun in a belt between two planets. Which **two** planets?

_____ and _____

[2 marks]

c What are asteroids? Put a tick (✓) in the box next to the correct answer.

clouds of very dense gas	
cubes of ice	
rocks that have broken away from planets	
rocks left over when the Solar System formed	

[1 mark]

2 The diagram shows the orbits of two bodies orbiting the Sun.
One is a planet, the other is a comet.

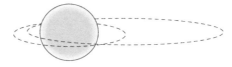

a Label the orbit of the comet with the letter **C**. [1 mark]

b When does the tail of a comet become visible?

_____ [1 mark]

c Why does a comet's tail always point away from the Sun?

_____ [1 mark]

3 Scientists are constantly updating information on the paths of **NEO**s.

a What are NEOs?

_____ [1 mark]

b Why is it important to constantly monitor the paths of NEOs?

_____

_____ [2 marks]

The Big Bang

1 Scientists think that 15 billion years ago there was a 'Big Bang'.

  a What was the Universe like before the Big Bang?

    _____ [1 mark]

  b What were three of the first things formed after the Big Bang?
    Put a (ring) around each of the **three** correct answers.

    copper   helium   hydrogen   iron   protons   uranium

    [3 marks]

  c The Universe is expanding. Galaxies in the Universe are moving at different speeds.
    Which galaxies are moving the fastest?

    _____
    _____ [1 mark]

2 a New stars are being formed all the time.
    How do stars start to form?

    _____
    _____ [1 mark]

  b A star's life is determined by its size.
    i What happens to a medium-sized star, like our Sun, at the end of its life?

    _____
    _____
    _____
    _____
    _____
    _____ [4 marks]

    ii What happens to a very large star at the end of its life?

    _____
    _____
    _____
    _____
    _____
    _____ [4 marks]

P2 LIVING FOR THE FUTURE

P2 Revision checklist

- I can describe how energy from the Sun can be used for heating and producing electricity. ☐

- I can describe how generators produce electricity. ☐

- I can describe how electricity is distributed via the National Grid. ☐

- I know what fuels are used in power stations and some of their advantages and disadvantages. ☐

- I can calculate power and the cost of using an electrical appliance for a certain time. ☐

- I know how to measure radioactivity and why there is background radiation. ☐

- I can list some uses of alpha, beta and gamma sources and relate their use to their properties. ☐

- I can describe the Earth's magnetic field and its similarity to a bar magnet and a coil. ☐

- I can describe how the Moon was formed. ☐

- I know the names of the planets and their order from the Sun. ☐

- I can describe how we are exploring space through manned and unmanned spacecraft. ☐

- I know that there are bodies in space other than planets and moons. ☐

- I can describe asteroids and comets and know the importance of constantly checking NEOs. ☐

- I know that scientists believe the Universe started with a Big Bang and that it is still expanding. ☐

P2 LIVING FOR THE FUTURE

Speed

1 a Melissa and Eve are running in a 1500 m race. Melissa finishes in 300 s. Eve takes 250 s.

   i Who runs faster, Melissa or Eve? _____ [1 mark]

   ii How do you know?

   _____ [1 mark]

b Ivy is going to measure the average speed of a trolley across the laboratory floor.

   i What measurements will she need to take?

   _____
   _____ [2 marks]

   ii Suggest what instruments she could use to obtain her measurements.

   _____
   _____ [2 marks]

c Holly's father got a speeding ticket. His car was photographed twice as it crossed white lines in the road 1.5 m apart. He passed over eight lines in the 0.5 s between each photograph. How fast was he travelling?

_____
_____
_____
_____ [4 marks]

d Freddie is on holiday in France. He travels 390 km in 3 hours on a French motorway.

   i Calculate the average speed of his car on this journey in km/h.

   _____
   _____ [3 marks]

   ii Why is your answer the **average** speed of the car?

   _____ [1 mark]

   iii The speed limit on the motorway is 130 km/h. Did the car break the speed limit? Explain your answer.

   _____
   _____ [2 marks]

2 The graph shows Ashna's walk to her local shop and home again.

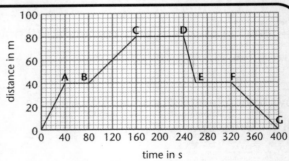

a Between which points did Ashna wait at traffic lights to cross the road?

_____ [1 mark]

b How long did she spend in the shop? _____ [1 mark]

c How far is the shop from her home? _____ [1 mark]

d During which part of her journey did she walk fastest? _____ [1 mark]

Changing speed

1 a Darren is riding his bicycle along a road. The speed–time graph shows how his speed changed during the first minute of his journey.

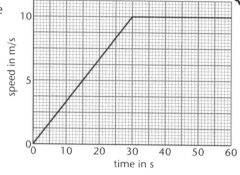

   i Describe how the speed changed, giving as much detail as possible.

_____

_____

_____

_____

_____ [4 marks]

   ii Darren makes the same journey the next day but:
- increases his speed at a steady rate for the first 20 s, reaching a speed of 10 m/s
- travels at a constant speed for 10 s
- slows down at a steady rate for 15 s to a speed of 5 m/s
- travels at a constant speed of 5 m/s.

Plot the graph of this journey on the axes given. [4 marks]

   iii How could you calculate the distance Darren travelled in the first minute of his original journey?

_____ [1 mark]

b The way in which the speed of a car changes over a 60 s period is shown in the table.

   i Plot a speed–time graph for the car using the axes given.

time in s	speed in m/s
0	0
5	5
10	10
15	15
20	15
25	15
30	15
35	15
40	15
45	11
50	7.5
55	3.5
60	0

[5 marks]

   ii The car is in an area where the speed limit is 50 km/h.
Does the driver exceed this limit? Show clearly how you decide.

_____

_____

_____

_____ [4 marks]

2 A car accelerates from 10 to 40 m/s in 6 s. Calculate its acceleration.

_____

_____

_____ [4 marks]

Forces and motion

1 The engine has just been switched on in the car shown in the diagram. P is the forward force on the car due to the pull of the engine.

a Put a ring around the word that **best** describes the motion of the car as it starts to move.

    accelerate   constant speed   decelerate   [1 mark]

b i How will the size of force P change if the driver presses the accelerator pedal harder?

_____ [1 mark]

  ii How will the motion of the car change?

_____ [1 mark]

c On another day the car carries four people and a fully loaded boot so its mass increases. How will its motion change if the value of P is the same as in **a**?

_____ [1 mark]

---

2 A car of mass 500 kg accelerates steadily from 0 to 40 m/s in 20 s.

a What is its acceleration?

_____

_____ [4 marks]

b What resultant force produces this acceleration?

_____

_____ [3 marks]

c The actual force required will be greater than your answer to **b**. Why?

_____ [1 mark]

---

3 Helen is driving her car on a busy road when the car in front brakes suddenly. She puts her foot firmly on her brake pedal and just manages to stop without hitting the car in front.

a What is meant by 'braking distance'?

_____ [1 mark]

b What is meant by 'thinking distance'?

_____ [1 mark]

c How can Helen's stopping distance be calculated?

_____ [1 mark]

d Later that day Helen is driving at high speed on a motorway. How will Helen's thinking distance change?

_____ [1 mark]

Explain why.

_____ [2 marks]

e Write down **two** things, apart from speed, that could increase a driver's thinking distance.

_____

_____ [2 marks]

Work and power

1 Hilary lifts a parcel of weight 80 N onto a shelf 2 m above the ground.

   a Would she do more or less work if the parcel weighed 120 N? _____ [1 mark]

   b Would she do more or less work if she lifted the parcel onto a shelf 1.6 m high?

      _____ [1 mark]

   c Hilary tries to lift a parcel weighing 500 N but she cannot move it. How much work does she do?

      _____ [1 mark]

   d Where does Hilary get the energy she needs to do work to lift the parcels?

      _____ [1 mark]

   e Calculate the amount of work Hilary does when she lifts the parcel of weight 80 N onto a shelf 2 m above the ground.

      _____

      _____ [3 marks]

2 Chris and Abi both have a mass of 60 kg. They both run up a flight of stairs 3 m high. Chris takes 8 s and Abi takes 12 s.

   a What can you say about the amount of work that each does? _____ [1 mark]

   b What can you say about the power of Chris and Abi?

      _____ [1 mark]

   c Calculate Chris' weight. [Take g = 10 N/kg.]

      _____ [2 marks]

   d How much work does Chris do in running up the stairs?

      _____

      _____ [3 marks]

   e Calculate Chris' power.

      _____

      _____

      _____ [3 marks]

   f Calculate Abi's power.

      _____

      _____ [2 marks]

3 The table shows the fuel consumption of three cars in miles per gallon (mpg).

car	fuel consumption (mpg)
A	48
B	34
C	28

   a Which car has the best fuel consumption? _____ [1 mark]

   b Which car is likely to be most powerful? _____ [1 mark]

   c We should keep our fuel consumption to a minimum to protect the environment. Why?

      _____

      _____ [3 marks]

Energy on the move

1 a What is meant by **kinetic energy**?
_____ [1 mark]

b Which possesses more kinetic energy?

	mass in kg	speed in m/s		mass in kg	speed in m/s
A	2	5	B	2	7
C	20	2	D	15	2

A or B _____   C or D _____   [2 marks]

2 a Use the data about the fuel consumption of petrol and diesel cars to answer the following questions.

engine size in litres	fuel consumption in mpg	
	petrol	diesel
1.6	44	60
2.0	40	51

  i Which car has the best fuel consumption? _____ [1 mark]

  ii Which type of fuel, petrol or diesel, is more efficient? _____ [1 mark]

  iii How did you decide?
  _____ [1 mark]

b Use the data about fuel consumption to answer the following questions.

car	fuel	engine size in litres	miles per gallon	
			urban	non-urban
Renault Megane	petrol	2.0	25	32
Land Rover	petrol	4.2	14	24

  i Suggest why fuel consumption is better in non-urban conditions.
  _____
  _____ [2 marks]

  ii How many gallons of petrol would a Land Rover use on a non-urban journey of 96 miles?
  _____ [2 marks]

  iii Would a Renault Megane use more or less fuel for the same journey? _____ [1 mark]

  iv Suggest a reason for the difference.
  _____ [1 mark]

3 a How do battery-powered cars pollute the environment?
_____
_____ [2 marks]

b i Give **one** advantage of solar-powered cars compared to battery-powered cars.
  _____ [1 mark]

  ii Give **one** disadvantage of solar-powered cars compared to battery-powered cars.
  _____ [1 mark]

P3 FORCES FOR TRANSPORT

Crumple zones

1 Kevin was involved in an accident on the M1 motorway. Luckily he was not seriously hurt but his car was badly damaged. Kevin was wearing a seat belt and his car had several safety features.

  a Which safety feature

    i badly damaged his car but helped to protect Kevin?

    _____ [1 mark]

    ii inflated and squashed to stop the steering wheel pushing into Kevin's chest?

    _____ [1 mark]

    iii stopped Kevin being thrown through the windscreen?

    _____ [1 mark]

  b How does each feature in the table help to reduce Kevin's injuries by absorbing energy?

safety feature	how it works
seatbelt	
crumple zones	
air bag	

[3 marks]

  c **i** What is meant by an 'active safety feature' on a car?

    _____ [2 marks]

    ii What is meant by a 'passive safety feature' on a car?

    _____
    _____ [2 marks]

  d **i** Give **two** examples of active safety features.

    _____ [2 marks]

    ii Give **two** examples of passive safety features.

    _____
    _____ [2 marks]

  e **i** Why should all safety features in a car be checked regularly?

    _____ [1 mark]

    ii Why is it essential to do this after a car crash?

    _____ [1 mark]

    iii Why must seatbelts always be replaced after a car crash?

    _____ [1 mark]

  f **i** What does **ABS** stand for?

    _____ [1 mark]

    ii Why are ABS brakes safer when a driver has to slam his foot on the brakes to stop quickly?

    _____
    _____ [2 marks]

Falling safely

1 Charlie drops a golf ball and a ping pong ball from a height of 30 cm above a table.

a Why do the balls fall towards the table?

_____ [1 mark]

b How does the speed of the balls change as they fall?

_____ [1 mark]

c Both balls hit the table together although their masses are different.
Charlie now drops the two balls from a height of 130 cm above the table.
Explain why the golf ball hit the table before the ping pong ball.

_____
_____ [2 marks]

d Next Charlie drops the golf ball and a feather from a height of 50 cm above the table. Which hits the table first? Put a (ring) around the correct answer.

     golf ball   feather   they arrive together   [1 mark]

e Sarah is a sky diver. She has a mass of 60 kg.

  i What is the value of her acceleration just after leaving the aircraft?

_____ [1 mark]

  ii On the diagram mark and name the forces acting on Sarah as she falls. [2 marks]

  iii What can you say about the size of these forces?

_____ [1 mark]

  iv Sarah's acceleration decreases as she falls. Explain why.

_____ [1 mark]

  v Eventually she is travelling at a constant speed. What is this speed called?

_____ [1 mark]

  vi What can you say about the size of the forces acting on her now?

_____ [2 marks]

2 Racing cyclists try to streamline their shape.

a Why do they do this?

_____
_____ [2 marks]

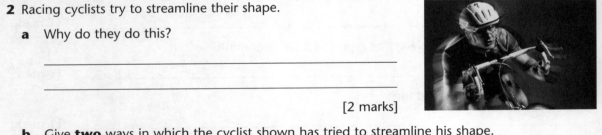

b Give **two** ways in which the cyclist shown has tried to streamline his shape.

_____
_____ [2 marks]

The energy of games and theme rides

Grades

P3 FORCES FOR TRANSPORT

G–E

1 Finish the sentences. Choose words from this list.

gravitational potential energy (GPE) **kinetic energy (KE)**
more **less**

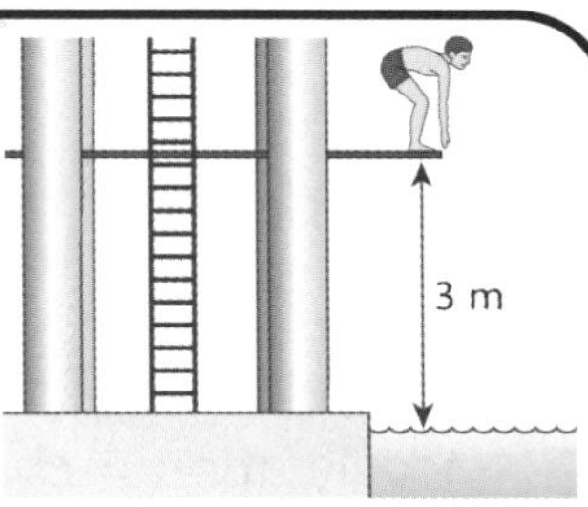

Rob is about to dive into the swimming pool.

He has ______________________ . As he dives ______________________ changes to ______________________ . Rob climbs to the 10 m board.

He has more ______________________ than before. [4 marks]

2 Kate is bouncing a ball. She drops it from **A** and it rises to **D** after the first bounce.

G–E

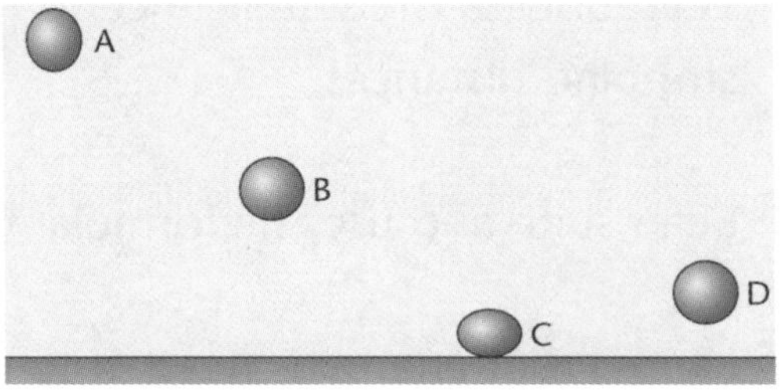

a What sort(s) of energy does the ball have at

A ______________________ [1 mark]

B ______________________ [1 mark]

C ______________________ [1 mark]

D–C

b Why is **D** much lower than **A**? ______________________

______________________ [2 marks]

D–C

3 The diagram shows a roller coaster. The carriages are pulled up to **B** by a motor and then released.

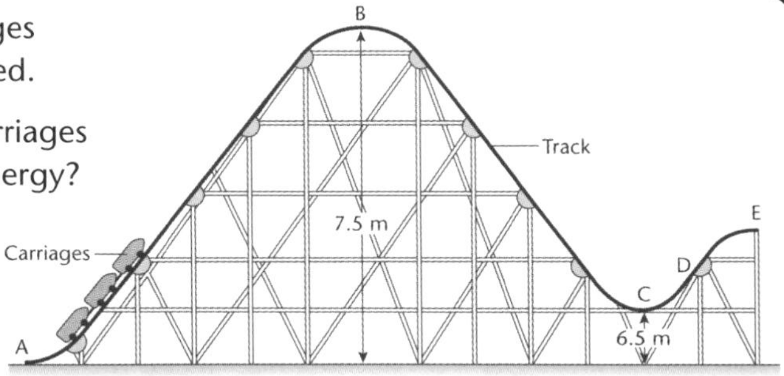

a At which point, **A**, **B**, **C**, **D** or **E**, do the carriages have the greatest gravitational potential energy?

______________________ [1 mark]

b How does the gravitational potential energy at this point change when the carriages are full of people?

______________________ [1 mark]

c At which point, **A**, **B**, **C**, **D** or **E** do the carriages have the greatest kinetic energy?

______________________ [1 mark]

d Describe the main energy change as the carriages move from **B** towards **C**.

______________________ [2 marks]

e Why must the height of the next peak at **D** be less than that at **B**?

____

______________________ [2 marks]

f The theme park decides to build a faster roller coaster. Suggest how they could modify the design to achieve this, using your knowledge of energy transfers.

____

____

______________________ [3 marks]

P3 Revision checklist

P3 FORCES FOR TRANSPORT

- I know that speed is measured in m/s and can use the formula: speed = distance ÷ time. ☐

- I can describe, draw and interpret distance–time graphs and speed–time graphs. ☐

- I know that acceleration is measured in m/s^2 and that: acceleration = change in speed ÷ time taken. ☐

- I can state and use the formula: force = mass x acceleration. ☐

- I can discuss the significance to road safety of thinking, braking and stopping distances. ☐

- I can state and use the formula: work done = force x distance. ☐

- I know that energy is needed to do work and that both work and energy are measured in joules (J). ☐

- I can state that power is measured in watts (W) and use the formula: power = work done ÷ time. ☐

- I can recognise objects that have kinetic energy (KE) and know the factors that increase it. ☐

- I can interpret data about fuel consumption. ☐

- I can describe typical safety features in modern cars. ☐

- I can describe how the motion of a falling object changes due to the effect of air resistance. ☐

- I can recognise objects that have gravitational potential energy (GPE). ☐

- I can interpret a gravity ride (roller coaster) in terms of GPE, KE and energy transfer. ☐

Sparks!

1 Sonya rubs a polythene strip with a cloth. The strip becomes charged. It can pick up small pieces of paper.

   a What charge does the polythene strip gain? _____ [1 mark]

   b **i** What material could she rub to obtain a strip with a different charge?

        _____ [1 mark]

     ii What effect will this strip have on the small pieces of paper?

        _____ [1 mark]

   c Ed rubs a copper strip with a cloth. Explain why it has no effect on the small pieces of paper.

       _____
       _____ [2 marks]

2 Sally stands on an insulating mat. She puts her hands on the dome of an uncharged Van de Graaff generator. The Van de Graaff generator is switched on and Sally's hair starts to stand on end.

   a Why does Sally stand on an insulating mat?

       _____ [1 mark]

   b Why must the Van de Graaff generator be uncharged when Sally puts her hands on it?

       _____ [1 mark]

   c **i** What happens to Sally when the Van de Graaff generator is switched on?

        _____ [1 mark]

     ii Why does this make Sally's hair stand on end?

        _____
        _____ [2 marks]

3 Use your knowledge of electrostatics to explain the following.

   a You sometimes get an electric shock on closing a car door after a journey.

       _____
       _____ [2 marks]

   b You should never shelter under a tree during a thunderstorm.

       _____ [1 mark]

   c Cling film often sticks to itself as it is unrolled.

       _____ [1 mark]

   d Priya becomes charged when she walks on a nylon carpet.

       _____
       _____ [2 marks]

   e Tom touched a bare wire at the back of his TV, when switched on, and got a serious shock.

       _____
       _____ [2 marks]

Uses of electrostatics

1 A defibrillator delivers an electric shock through the chest wall to the heart.

  a What is the purpose of defibrillation? _____

      [1 mark]

  b What does the electric shock do to the heart?

    _____

      [1 mark]

  c The paddles, charged from a high voltage supply, are placed on the patient's chest. How does the operator ensure that there is good electrical contact with the patient's chest?

    _____

    _____

      [2 marks]

  d A current of about 50 A passes through the patient for about 4 ms (0.004 s). In general, such a large current would be fatal.
Why can it be used in this situation?

      [1 mark]

2 In a bicycle factory the frames are painted using an electrostatic sprayer. The paint is positively charged. The frames are given the opposite charge to the paint.

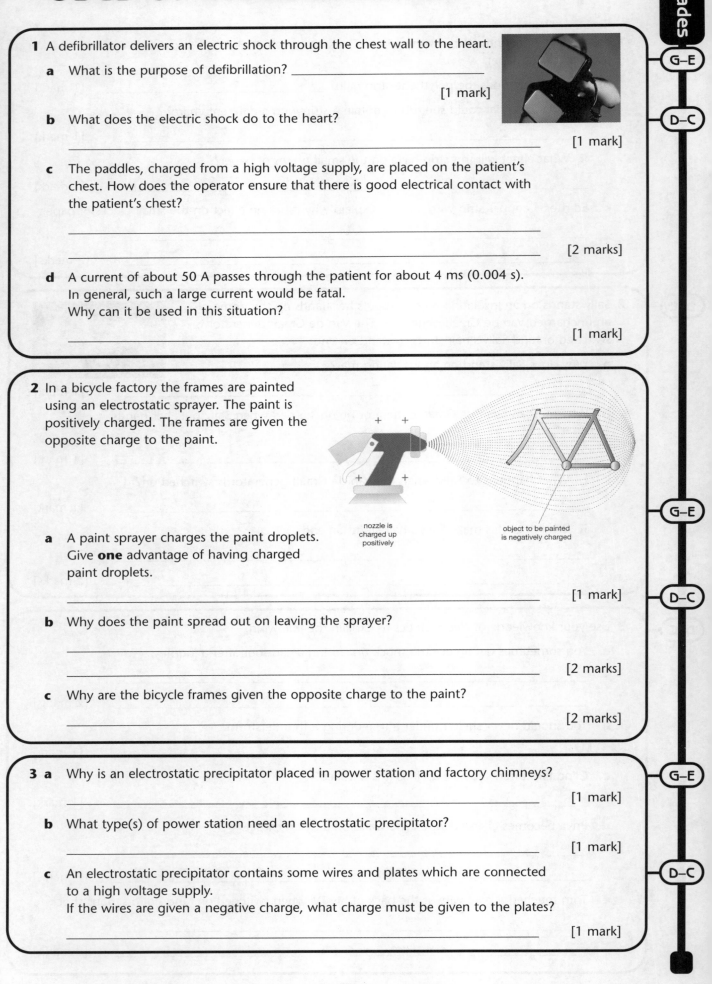

nozzle is charged up positively

object to be painted is negatively charged

  a A paint sprayer charges the paint droplets. Give **one** advantage of having charged paint droplets.

    _____

      [1 mark]

  b Why does the paint spread out on leaving the sprayer?

    _____

    _____

      [2 marks]

  c Why are the bicycle frames given the opposite charge to the paint?

    _____

      [2 marks]

3 a Why is an electrostatic precipitator placed in power station and factory chimneys?

    _____

      [1 mark]

  b What type(s) of power station need an electrostatic precipitator?

    _____

      [1 mark]

  c An electrostatic precipitator contains some wires and plates which are connected to a high voltage supply.
If the wires are given a negative charge, what charge must be given to the plates?

    _____

      [1 mark]

Safe electricals

1 Lee sets up the circuit shown. The lamp does not light.

  a **i** Why does the lamp not light? _____ [1 mark]

    ii Change Lee's circuit so that the lamp will light. [1 mark]

  b Meera sets up another circuit.
    i What effect will this have on the brightness of the lamp?

    _____ [1 mark]

    ii Add an ammeter and voltmeter to Meera's circuit to allow her to measure the current through the lamp and the potential difference across the lamp. [2 marks]

    iii If the voltmeter reads 6 V and the ammeter 0.25 A, calculate the resistance of the lamp.

    _____

    _____ [4 marks]

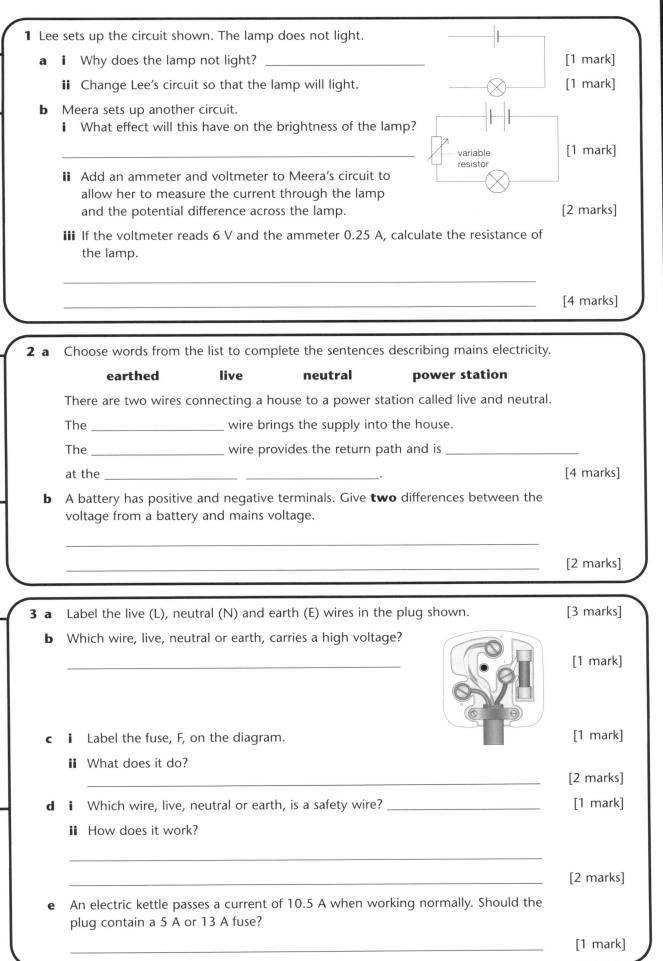

2 a Choose words from the list to complete the sentences describing mains electricity.

    earthed   **live**   **neutral**   **power station**

  There are two wires connecting a house to a power station called live and neutral.

  The _____ wire brings the supply into the house.

  The _____ wire provides the return path and is _____

  at the _____ _____. [4 marks]

b A battery has positive and negative terminals. Give **two** differences between the voltage from a battery and mains voltage.

_____

_____ [2 marks]

3 a Label the live (L), neutral (N) and earth (E) wires in the plug shown. [3 marks]

  b Which wire, live, neutral or earth, carries a high voltage?

  _____ [1 mark]

  c **i** Label the fuse, F, on the diagram. [1 mark]

    ii What does it do?

    _____ [2 marks]

  d **i** Which wire, live, neutral or earth, is a safety wire? _____ [1 mark]

    ii How does it work?

    _____

    _____ [2 marks]

  e An electric kettle passes a current of 10.5 A when working normally. Should the plug contain a 5 A or 13 A fuse?

  _____ [1 mark]

Ultrasound

1 a What is meant by a **longitudinal** wave?

_____

_____ [2 marks]

b What is meant by the **frequency** of a wave?

_____ [1 mark]

c Sound is a longitudinal wave.

  i Explain how sound travels through the air to reach your ear.

_____

_____ [2 marks]

  ii How does the frequency of a note change if its pitch increases?

_____ [1 mark]

  iii What is 'ultrasound'?

_____ [1 mark]

2 a Why does a pregnant woman usually have an ultrasound scan?

_____ [1 mark]

b The ultrasound waves used vibrate 1 000 000 times a second. What is the frequency of the ultrasound?

_____ [1 mark]

c Give **two other** uses of ultrasound.

_____

_____ [2 marks]

d Finish these sentences about an ultrasound scan. Choose words from this list.

    echoes   **gel**   **image**   **probe**   **pulse**

    reflected   **skin**   **tissues**   **ultrasound**

A _____ of ultrasound is sent into a patient's body. At each boundary between different _____ some ultrasound is _____ and the rest is transmitted.

The returning _____ are used to build up an _____ of the internal structure.

A _____ is placed on the patient's body between the ultrasound _____ and their _____. Without it nearly all the _____ would be _____ by the _____.

[4 marks]

e **High-powered** ultrasound is used to treat a patient with kidney stones.

  i How does ultrasound do this?

_____

_____ [3 marks]

  ii Why must **high-powered** ultrasound be used?

_____ [2 marks]

Treatment

1 a In medicine, what is the difference between diagnosis and therapy?

_____

_____ [2 marks]

b Give **one** similarity and **one** difference between X-rays and gamma rays.

_____

_____ [2 marks]

c Why are X-rays and gamma rays suitable to treat cancer patients?

_____

_____ [2 marks]

d Why are alpha and beta particles **not** suitable to treat cancer patients?

_____

_____ [2 marks]

2 a Why is nuclear radiation suitable for treating cancer?

_____ [1 mark]

b What is 'radiotherapy'?

_____ [1 mark]

c Why is radiotherapy often used after surgery to remove a cancerous tumour?

_____

_____ [2 marks]

d Give another use of gamma radiation in hospitals.

_____ [1 mark]

3 a What is a radioactive tracer?

_____

_____ [2 marks]

b Why is it used?

_____

_____ [2 marks]

c What sort of radiation should a tracer emit?

_____ [1 mark]

d Which organ of the body is investigated using iodine-123 as a tracer?

_____ [1 mark]

e X-rays and gamma rays have similar properties.
Why are gamma rays suitable to use as tracers but X-rays are not?

_____

_____ [1 mark]

What is radioactivity?

1. Jay is going to measure the activity of a radioactive source.

   a. What is meant by the 'activity' of a radioactive source?

   _____

   _____ [2 marks]

   b. i. What will he use to detect the radiation emitted by the source?

   _____ [1 mark]

   ii. What will he use to measure the rate at which radiation is emitted by the source?

   _____ [1 mark]

   c. What does each count recorded represent?

   _____ [1 mark]

   d. Jay records a count of 750 in 30 s.
      i. What is the activity of the source? Show how you work out your answer.

   _____

   _____

   _____

   _____ [4 marks]

   ii. Would you expect the activity of the source to increase or decrease as time passes?

   _____ [1 mark]

   iii. Jay's teacher tells him he should repeat his readings several times to find an average value for the activity of the source. Why?

   _____

   _____ [2 marks]

2. a. Finish the table about the **three** types of nuclear radiation.

type of radiation	charge (+, – or 0)	what it is	particle or wave
alpha			
beta			
gamma			

[5 marks]

   b. Name the type of nuclear radiation that

   is the most penetrating _____

   is stopped by several sheets of paper _____

   has the greatest mass _____

   does not change the composition of the nucleus _____

   travels at about one-tenth the speed of light _____

   [5 marks]

Uses of radioisotopes

1 a What is meant by background radiation?

_____

_____ [2 marks]

b Suggest **two natural** sources of background radiation.

_____

_____ [2 marks]

2 Write down **three** examples of the use of a radioisotope as a tracer.

_____

_____

_____ [3 marks]

3 Smoke alarms use a source of alpha radiation in a small chamber.

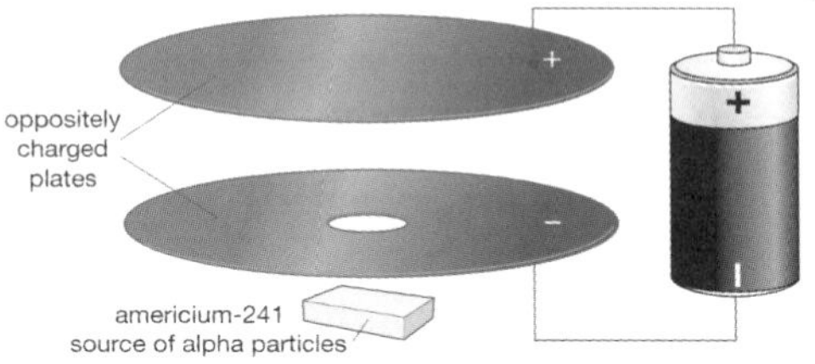

a Why is alpha radiation more suitable than either gamma or beta radiation for use in a smoke alarm?

_____ [1 mark]

b Explain how the smoke alarm works.

_____

_____

_____

_____ [4 marks]

4 Trees contain carbon-14 which is radioactive. The graph shows how the activity of 1 kg of wood changes after a tree has died. This can be used to estimate the age of objects that were once alive.

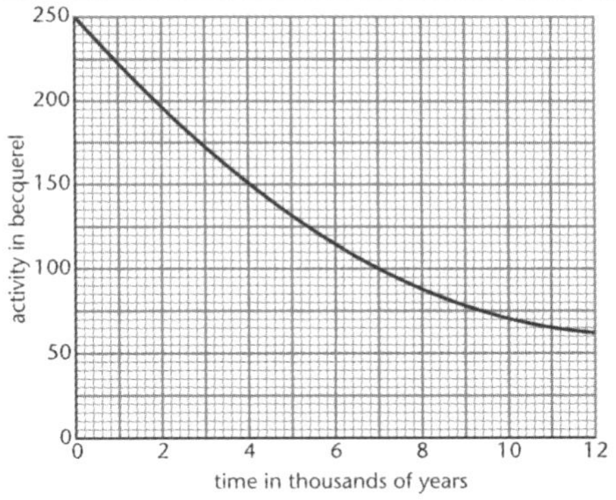

a This method cannot be used to date a wooden bowl made from a tree that died less than 200 years ago. Explain why.

_____

_____ [1 mark]

b The remains of an ancient civilisation were found. Put a ring around objects that could be dated using this method.

    animal bone   **bronze tool**   **seeds**   **stone**   [2 marks]

c Why cannot carbon-14 be used to date rocks?

_____ [1 mark]

Fission

1 A power station makes electricity.

  a Label the diagram.

    generator, source of energy, steam, water, turbine

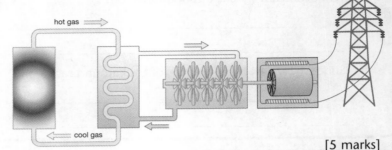

[5 marks]

  b Finish the sentences to explain how a power station works. Choose words from the list used to label the diagram.

    The _____ provides heat to boil the _____ to produce _____. The pressure of the _____ turns the _____ which turns the _____ making electricity. [6 marks]

2 a i What element is used as the fuel in a nuclear power station? _____ [1 mark]

    ii What is meant by 'fission'?

    _____ [2 marks]

  b i What do we call the process that allows fission to continue until all the fuel is used up?

    _____ [1 mark]

    ii What is the difference between a nuclear reactor in a power station and a nuclear bomb?

    _____ [2 marks]

3 a i How can materials be made radioactive?

    _____ [1 mark]

    ii Give one use of a man-made radioisotope.

    _____ [1 mark]

  b i What is a neutron?

    _____ [2 marks]

    ii Radioisotopes can be produced by bombarding atoms with neutrons. Why are neutrons good at doing this?

    _____ [3 marks]

4 Nuclear power stations produce radioactive waste.

  a Suggest **one** method of disposing of low level radioactive waste.

    _____ [1 mark]

  b Suggest **one** method of disposing of high level radioactive waste such as spent fuel rods from a nuclear reactor.

    _____ [1 mark]

P4 Revision checklist

- I know that there are two kinds of electric charge, positive and negative. ☐

- I can explain how static electricity can sometimes be dangerous and sometimes a nuisance. ☐

- I can describe some uses of static electricity. ☐

- I can explain the behaviour of simple circuits and how resistors are used in circuits. ☐

- I can state and use the formula: resistance = voltage ÷ current. ☐

- I know about live, neutral and earth wires, fuses, circuit breakers and double insulation. ☐

- I can describe the key features of longitudinal waves. ☐

- I know about ultrasound and can describe some medical uses of it. ☐

- I can describe how nuclear radiation is used in hospitals. ☐

- I can describe the properties of nuclear radiation. ☐

- I can state what alpha and beta particles are. ☐

- I can describe background radiation and state what causes it. ☐

- I can describe some non-medical uses of radioisotopes. ☐

- I can describe how domestic electricity is generated in a nuclear power station. ☐

P4 RADIATION FOR LIFE

Satellites, gravity and circular motion

1 a Write down the name of the Earth's natural satellite.

_____ [1 mark]

b The Sun has many satellites in orbit around it.

What special name do we call satellites that orbit the Sun?

_____ [1 mark]

c A television station transmits a signal to a satellite. What happens to the signal when it reaches the satellite? Finish the sentence.

The signal is _____ and _____ [1 mark]

d How long does it take a geostationary satellite to orbit the Earth?

_____ [1 mark]

e What is the most common use for geostationary satellites?

_____ [1 mark]

f Polar orbiting satellites are used to help forecast the weather. Why are polar orbiting satellites used?

_____ [1 mark]

2 a STENSAT is a communications satellite orbiting the Earth. Write down the name of the centripetal force that keeps STENSAT in orbit.

_____ [1 mark]

b What is meant by a **centripetal** force?

_____ [2 marks]

Vectors and equations of motion

1 a Dave is driving his car at 60 mph. He overtakes another car travelling at 40 mph. What is Dave's speed relative to the other car? Put a ring around the correct answer.

20 mph   40 mph   50 mph   60 mph   100 mph

[1 mark]

b What is the difference between a scalar quantity and a vector quantity?

_____
_____ [2 marks]

c David and Jean are both pushing their father's car in the same direction. David pushes with a force of 600 N and Jean with a force of 450 N. What is the resultant force on the car?

_____ [1 mark]

d John is trying to open the door. He pushes with a force of 550 N. His friends, Jean and Erica are trying to stop him from opening the door. Jean pushes with a force of 300 N and Erica with a force of 250 N. What is the resultant force on the door?

_____ [1 mark]

2 The maximum speed of a car on a motorway is 70 mph. Tina drove for 3 hours and was surprised to find she had only travelled 186 miles. She was sure that whenever she looked at her speedometer, it was reading 70 mph.

a Suggest why she had only travelled 186 miles instead of 210 miles.

_____ [1 mark]

b Calculate her average speed.

_____
_____
_____
_____ [3 marks]

c When Tina joined the motorway, she was travelling at 9 m/s as she accelerated along the slip road. Her acceleration was 3.6 m/s². She accelerated for 5 s until she reached the end of the slip road. Calculate her final velocity.

_____
_____
_____
_____ [3 marks]

Projectile motion

1 a Finish the sentences by choosing the **best** words from this list.

ellipse   football   parabola   projectile   trajectory

Any object that moves in the Earth's gravitational field is called a _____.

The path the object takes is known as its _____.

The shape of this curved path is a _____. [3 marks]

b Mark throws a ball horizontally as hard as he can from the top of a tall cliff. At the same time, Mary drops a similar ball from the same height. It takes 2.4 s for Mary's ball to hit the sea below.

  i How long does it take for Mark's ball to hit the sea?

  Put a (ring) around the correct answer.

  less than 2.4 s

  exactly 2.4 s

  more than 2.4 s [1 mark]

  ii Mark throws the ball with a horizontal velocity of 4 m/s. What is the **horizontal** velocity when the ball hits the sea?

  _____ [1 mark]

c Nathan fires an arrow at a target. It falls just short. He cannot pull the bow string with any more force. What can he do to increase the range of his arrow?

_____ [1 mark]

2 Newton suggested that if you could throw a ball with enough speed from the top of a tall tower, it would never come back to Earth but would stay in orbit. Explain why the ball would not come back to Earth.

_____
_____ [2 marks]

Momentum

1 a Finish the sentences by choosing the **best** words or numbers from this list.

action   momentum   reaction   resultant   0   60   120

Gita and Shawna are on roller blades facing one another.

Shawna **pushes** Gita with a force of 60 N.

The force Shawna uses is called the _____.

Gita exerts a force on Shawna. This force is called the _____.

The size of the force Gita exerts on Shawna is _____ N. [3 marks]

b An apple has a mass of 100 g. Its **weight** is 1 N. Why does the apple have a weight?

_____ [1 mark]

2 a The speed limit outside many schools is 20 mph. This is to reduce injuries if there is a **collision** involving a car. What is meant by the word collision?

_____ [1 mark]

b Gary's mass is 75 kg. He is in a car travelling at 9 m/s.

  i Calculate Gary's momentum.

  _____
  _____ [3 marks]

  ii Gary is driving past a school when his car skids and collides with a solid stone wall. Gary is wearing his seat belt so he is not injured. The seat belt means it takes 0.5 s for Gary to stop moving after the collision. Calculate the force on Gary.

  _____
  _____ [2 marks]

  iii If Gary had not been wearing his seat belt, it would only have taken 0.05 s for Gary to stop as he hit the steering wheel. How would this affect the force on him?

  _____ [2 marks]

  iv 9 m/s is about 20 mph. How would the force on Gary differ if he had been speeding past the school at 60 mph when he collided with the wall?

  _____
  _____ [2 marks]

Satellite communication

1 a Two of these wavelengths are **not** wavelengths for radio waves. Put (rings) around the two wavelengths which are not radio wavelengths.

1 mm   **1 cm**   **1 m**   **1 km**   **1000 km**   **1 000 000 km**

[2 marks]

b What **two** things may happen to low frequency radio waves when they pass into the Earth's atmosphere. Choose from:

absorbed   **diffracted**   **reflected**   **transmitted**

_____ and _____ . [2 marks]

c Most televisions have an **aerial**. What is the job of the aerial?

_____ [1 mark]

d Some televisions receive their signals from a geostationary satellite. These televisions have a different type of receiver. What do satellite televisions use to receive the signal?

_____ [1 mark]

e A geostationary satellite is used for communications. A television company transmits a microwave signal from its Earth station to a satellite. What happens to this signal once it is received by the satellite?

_____
_____ [2 marks]

2 Microwaves do not **diffract** as well as long-wave radio waves.

a Explain the difference between microwaves and long-wave radio signals.

_____ [1 mark]

b Describe what is meant by the word 'diffraction'. You may find it easier to draw a diagram.

[1 mark]

Nature of waves

1 a Finish the sentences by choosing the **best** words from this list.

brighter   darker   longer   louder   quieter

When two sound waves interfere and add together, the sound gets _____.

When two light waves interfere and add together, the light gets _____.

When two sound waves interfere and subtract from one another, the sound gets _____.

When two light waves interfere and subtract from one another, the light gets _____. [4 marks]

b Ray connects two loudspeakers to the same sound generator. He places them about 2 m apart on the laboratory bench. He stands in front of one loudspeaker and walks towards the other. Describe what he hears as he walks from one loudspeaker to the other.

_____
_____
_____ [2 marks]

2 a The bar code readers in supermarkets are scanned by a special type of light. What is this light called? _____ [1 mark]

b What property of light allows bar code readers to work? Finish the sentence.

Light travels in _____ lines. [1 mark]

c Kirsty arranges the apparatus shown below.

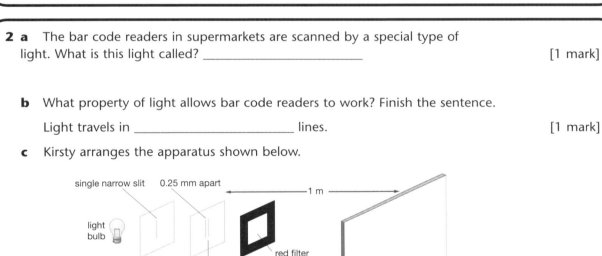

Describe what Kirsty sees on the screen.

_____
_____
_____ [2 marks]

d Duncan wants to demonstrate polarisation of sound waves. His teacher tells him this is impossible. Why can Duncan not show polarisation of sound waves?

_____ [2 marks]

Refraction of waves

1 a Light changes direction when passing from one material into another. What is this called? Put a (ring) around the correct answer.

diffraction   dispersion   polarisation   reflection   refraction [1 mark]

b Finish the sentence by choosing the **best** word from this list.

diffraction   dispersion   polarisation   reflection   refraction

When sunlight passes through raindrops to form a rainbow, this is known as

_____ [1 mark]

c Light travels at 300 000 km/s in air. What is its speed in glass? Put a (ring) around the correct answer.

20 m/s   200 m/s   200 000 km/s   300 000 km/s   400 000 km/s [1 mark]

d Use ideas about refractive index to explain why a rainbow forms.

_____
_____ [2 marks]

2 a Describe what happens to a ray of light as it passes from plastic into air with an angle of incidence of 85°. You may draw a diagram to illustrate your answer.

_____ [1 mark]

b Describe one use of optical fibres.

_____
_____ [2 marks]

Optics

1 a Finish the sentences by choosing the **best** words from this list.

concave   convex   focus   reflected   refracted

A lens that is fatter in the middle than at the edges is called

a _____ lens.

Light is _____ as it passes through the lens.

The light is converged to a _____. [3 marks]

b How can you tell whether or not the image produced by a lens is real?

_____ [1 mark]

c Both the camera and the projector use lenses to produce an image. What is the difference between the image produced by a camera and an image produced by a projector?

_____ [1 mark]

2 a Write down the name of the optical instrument that uses a convex lens placed close to an object to produce a much larger image.

_____ [1 mark]

b The film projector has several parts to it. Finish the table by writing in the job of each part.

part	job
curved mirror	
condenser lenses	
projection lens	

[3 marks]

P5 Revision checklist

- I know that a gravitational force keeps satellites in orbit. ☐

- I know about different satellite orbits and the uses of satellites in those orbits. ☐

- I can calculate the relative speed of two objects moving in the same straight line. ☐

- I know the difference between scalar and vector quantities. ☐

- I can recognise a projectile and the path it takes as it moves through the air. ☐

- I understand that a projectile has constant horizontal velocity but accelerates vertically. ☐

- I know that action and reaction are equal and opposite. ☐

- I can calculate the momentum of a moving object. ☐

- I understand the importance of momentum in the design of safe cars. ☐

- I know that the behaviour of radio waves in the Earth's atmosphere depends on their frequency. ☐

- I can describe diffraction as the spreading out of waves as they pass through a gap or round an obstacle. ☐

- I know the effects of sound and light interfering. ☐

- I can explain the effect of waves reinforcing and cancelling. ☐

- I can describe what happens when light passes from one material into another. ☐

- I know that refraction occurs because of a change in speed of a wave as it passes from one material into another. ☐

- I can describe the shape of a convex lens and list some of its uses. ☐

- I can describe the effect of a convex lens on a beam of light. ☐

Resisting

1 a Finish the table by writing in the quantity or unit.

quantity	unit
current	
	ohm
voltage	

[3 marks]

b Colin has dimmer switches in his house. They control the brightness of the bulbs. How does the resistance in the circuit change if he adjusts the dimmer switch to make the bulbs dimmer?

_____ [1 mark]

c A bulb has the figures 230 V, 0.5 A printed on it. Calculate the resistance of the bulb.

_____

_____ [2 marks]

2 a Draw a circuit diagram to show a voltmeter placed in parallel and an ammeter placed in series with a resistor.

[2 marks]

b Finish the sentences by choosing the **best** words from this list.

cold   decrease   hot   increase   stay the same

When a current passes through a wire, the wire

gets _____. This causes the resistance of the

wire to _____. [2 marks]

c Sketch a graph to show how the voltage across a bulb changes as the current is increased.

[2 marks]

Sharing

1 a Finish the sentences by choosing the **best** words from this list.

current   divider   multiplier   resistance   voltage

The potential difference across a resistor is also called its _____.

The _____ of a wire increases if its length increases.

The volume control of a hi-fi unit contains a potential _____. [1 mark]

b The diagram shows a potential divider circuit.

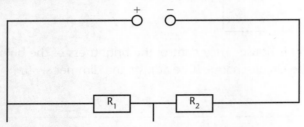

The output from the circuit is fixed. What change can be made to the circuit to allow the output to be variable?

_____ [1 mark]

2 a What causes the resistance of a thermistor to change?

_____ [1 mark]

b How is the behaviour of a thermistor different to the behaviour of a resistor?

_____ [1 mark]

c Explain why street lights often have light-dependent resistors on top of them.

_____

_____ [2 marks]

Motoring

1 a A wire is placed between the poles of a magnet. When a current is passed along the wire, the wire moves. Why?

[3 marks]

b What happens to the wire if the current direction is reversed?

[1 mark]

c What happens to the wire if the poles of the magnet are reversed?

[1 mark]

d Draw a diagram to show the magnetic field around a straight wire. You must show the direction of the current and the direction of the field.

[2 marks]

2 a Describe **one** use of an electric motor in the home.

[1 mark]

b A simple electric motor consists of a coil of wire rotating in a magnetic field. What **three** things can be done to make the motor spin faster?

[3 marks]

Generating

1 Marcus builds a model motor. He connects the motor to an ammeter. He spins the coil and a direct current is produced.

   a What type of current is produced by the generator at a power station?

      _____
      [1 mark]

   b Marcus now connects the ends of a length of wire to the ammeter. He holds the wire between the poles of a magnet. What does he see on the ammeter?

      _____
      [1 mark]

   c He moves the wire upwards out of the gap between the poles of the magnet. What does he see on the ammeter?

      _____
      [1 mark]

   d He moves the wire downwards into the gap between the poles of the magnet. What does he see on the ammeter?

      _____
      [2 marks]

2 a The bicycle dynamo contains a permanent magnet rotating inside a coil. The magnet is turned by the wheel of the bicycle rotating. What happens to the bicycle lamp when the cyclist stops at traffic lights?

      _____
      [1 mark]

   b The magnetic field in the generator at a power station is not produced by a permanent magnet. Describe how the field is produced.

      _____
      _____
      [2 marks]

   c A turbine is used to turn the rotor coil. Why is it important that the rotor turns at an almost constant speed?

      _____
      _____
      [2 marks]

Transforming

1 a What is the job of a **step-up** transformer?

_____ [1 mark]

b Describe one use for an **isolating** transformer.

_____ [1 mark]

c Kate uses a transformer to make her door bell work from the mains supply.

  i What type of transformer is used?

  _____ [1 mark]

  ii The door bell works on 12 V. The number of turns on the primary coil of the transformer is 10 000. The number of turns on the secondary coil is 500. Calculate the mains voltage that must be supplied to the primary coil.

  _____
  _____
  _____ [3 marks]

2 Electricity is transmitted around the country along overhead power lines.

  a What is this overhead network called? _____ [1 mark]

  b What is the voltage used to transmit electricity around the country? Put a (ring) around the correct answer.

  12 V   120 V   230 V   1000 V   400 000 V [1 mark]

  c Finish the sentences by choosing the **best** words from this list.

  cold   eight   energy   fields   four   hot   magnetism   two

  When a current passes through the overhead wires, the wires get _____. This means that _____ is lost to the surroundings as heat. If the current in the wires is reduced by a factor of two, the loss to the surroundings is reduced by a factor of _____. [3 marks]

Charging

1 a What is the circuit symbol for a diode? Put a (ring) around the correct answer.

[1 mark]

b Draw a diagram to explain what happens during **half-wave rectification**.

[2 marks]

c Sketch a graph to show the current–voltage characteristics for a diode.

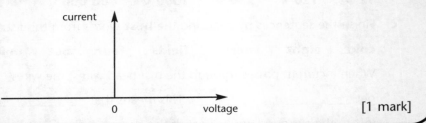

[1 mark]

2 a What is the job of a capacitor. Put a (ring) around the correct answer.

change AC into DC   **change DC into AC**

reverse direction of current   **store charge**

[1 mark]

b A capacitor is sometimes placed in a bridge rectifier circuit. What effect does this have on the output from the rectifier circuit?

_____

[1 mark]

It's logical

1 a Finish the sentences by choosing the **best** words or numbers from this list.

0   1   5   12   240   analogue   digital   high   low

A logic gate needs a voltage of about _____ V to make it work.

A signal is described as being _____ and is often represented by the number _____.

The absence of a signal is described as being _____ and is often represented by the number _____. [5 marks]

b A NOT gate is connected into a circuit. There is an input to the NOT gate. Describe any output.

_____ [1 mark]

c Look at the truth tables. What type of logic gate does each represent?

input 1	input 2	output
0	0	0
0	1	1
1	0	1
1	1	1

logic gate _____ [1 mark]

input 1	input 2	output
0	0	0
0	1	0
1	0	0
1	1	1

logic gate _____ [1 mark]

2 Draw a circuit diagram showing a potential divider circuit that can be used to switch on a logic gate in different temperature conditions.

[2 marks]

Even more logical

1 a What name is given to a combination of logic gates? Put a ring around the correct answer.

combine   lockgate   supergate   system

[1 mark]

b Helena connects the following logic circuit.

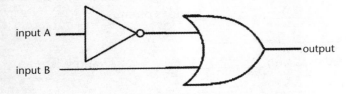

Construct the truth table for this circuit.

[4 marks]

2 Tammy connects a light-emitting diode (LED) to the output from a logic gate. It lights.

a She then connects a bulb to the same output. It does not light. Explain why.

_____ [1 mark]

b Suggest what device Tammy can use between the logic gate and the bulb that would allow the logic gate to control whether or not the bulb lights.

_____ [1 mark]

c Suggest **one** use for a light-emitting diode (LED) connected to the output from a logic gate.

_____ [1 mark]

P6 Revision checklist

- I know basic circuit symbols and electrical units. ☐
- I can explain the effect of using a variable resistor in a circuit. ☐
- I can use the equation: resistance = voltage ÷ current. ☐
- I know how to use a potential divider in a circuit. ☐
- I can describe the magnetic field patterns around a wire and different lengths of coil. ☐
- I can describe what happens when a current is passed through a coil in a magnetic field. ☐
- I know how to produce an electric current in a wire when moving it through a magnetic field. ☐
- I know that a dynamo is a motor working in reverse. ☐
- I can explain how a generator produces an alternating current. ☐
- I know that a transformer changes the size of a voltage. ☐
- I can identify where step-up, step-down and isolating transformers are used. ☐
- I can use and manipulate the equation: $V_p \div V_s = n_p \div n_s$. ☐
- I know why electricity is transmitted at high voltages ☐
- I can describe circuits for half-wave and full-wave rectification. ☐
- I can recognise the current–voltage characteristics for a diode. ☐
- I know why capacitors are used in electrical circuits. ☐
- I can describe, with the aid of truth tables, the behaviour of NOT, AND and OR gates. ☐
- I can describe how LDRs, thermistors and switches can be used in potential divider circuits to switch on logic gates under varying conditions. ☐
- I can explain why a relay is used at the output from a logic gate. ☐

P6 ELECTRICITY FOR GADGETS

Glossary

A

ABS Antilock braking system, which helps cars to stop safely.

absolute scale A scale linked to an external value, for example the absolute scale of temperature uses the value of −273 °C as zero.

accelerate To increase the velocity of an object.

acceleration The change of velocity in a given time, it is usually measured in metres per second squared (m/s^2).

air resistance The force air exerts on an object moving through it; air resistance slows down movement and increases as the moving object gets faster.

alpha particle A positively charged particle made of two protons and two neutrons emitted from the nucleus of a radioactive atom.

alternating current Current that rapidly reverses in direction. Mains electricity is supplied as alternating current or AC.

ammeter A meter that measures the current in a wire, in amps.

ampere The unit used to measure electrical current, often abbreviated to amp.

amplitude The maximum displacement of a wave from its rest position.

analogue A signal that shows a complete range of frequencies. Sound is analogue.

arbitrary scale A scale that only works within a given situation, for example, to say a value is twice a lower value without connecting either value to an external fixed point.

asteroids Lumps of rock orbiting the Sun; too small to be called planets, most asteroids are found in the area between Mars and Jupiter.

atom The smallest part of an element, atoms consist of negatively charged electrons around a positively charged nucleus.

B

background radiation Radiation from space and rocks; it is around us all the time and is at a very low level.

becquerel The unit of radioactivity, equal to one nuclear decay or transformation per second.

beta particle A negatively charged particle emitted from the nucleus of a radioactive atom.

Big Bang The event believed by many scientists to have been the start of the Universe.

braking distance The distance it takes to stop a moving vehicle after the brakes have been applied.

C

cancer A dangerous illness caused by radiation, smoking and some types of chemicals; cancer cells grow out of control to make lumps of cells and can invade normal healthy areas of the body.

capacitors Electronic components which can store electrical charge.

carbon dioxide A gas containing only carbon and oxygen; its chemical formula is CO_2.

carbon monoxide A poisonous gas containing only carbon and oxygen; its chemical formula is CO.

carbon-14 A radioactive isotope of carbon used in carbon-dating calculations.

carrier wave An electromagnetic wave that can carry speech, music, images, or other signals.

cataract An area of the lens or cornea becomes opaque so that light cannot pass through.

chain reaction A reaction where the products cause the reaction to go further or faster, e.g. in nuclear fission.

charge The property of matter that produces electrical interactions, charge can be positive or negative.

circuit The complete path around connected electrical components through which a current passes.

circular motion Movement in a circle.

comet A small body that orbits the Sun in a very elliptical orbit. Comets often show a tail caused by particles of dust and ice.

compression To push something together, to squeeze it and make it smaller.

conduction Movement of energy as heat (or electricity) through a substance without the substance itself moving.

conductor A substance that will let heat or electricity pass through it, e.g. copper.

constellation A group of stars that has been given a name by astronomers because they seem to be close together when seen from Earth. Members of a constellation may actually be many light years away from each other.

convection Movement of a heated substance which carries energy with it as heat.

convection current Movement of heated gases or liquids to float on top of the cooler, denser layers.

convex lens A lens that is thicker in the middle than at the edges; it bends light rays to a focus.

cosmic ray Radiation from space which hits the atmosphere. Some passes through while some is blocked. The northern lights are displays of light created when certain types of cosmic rays hit the magnetic field around the Earth.

count rate The number of nuclear events in a given time, often measured by a Geiger counter.

critical angle The angle above which light is reflected back into a glass block.

current The flow of electrical charge through an electrical circuit.

D

decay The emission of nuclear radiation in the form of alpha, beta and gamma radiation.

deceleration Reduction of the velocity of a moving object; it is usually calculated as change in velocity per second and given a minus sign.

diffraction Spreading of waves after passing through a narrow slit or around an object.

dimmer switches Switches that contain a variable resistor which can dim lights.

diodes An electronic component that only lets current pass through it in one direction.

dispersion The separation of white light into a coloured spectrum after refraction.

double-insulated An electrical device in which there are at least two layers of insulation between the user and the electrical wires.

drag A force which slows down something moving through a liquid or a gas.

dust precipitator A device that uses electrostatic charge to remove dust from the air.

dynamo A device that converts movement energy into electricity.

E

earthed To be connected to the earth – an important safety feature of many electrical appliances.

efficient An efficient device transfers most of the input energy into the desired output energy.

electromagnetic wave A transverse wave with electrical and magnetic properties which can pass through a vacuum.

electron A small negatively charged particle that orbits around the nucleus of an atom.

electrostatic To do with electric charges that are not moving.

elliptical A path that follows an ellipse – which looks a bit like a flattened circle.

endoscope A device using optical fibres to look inside another object, typically used by doctors to look inside the stomach or colon.

energy Energy is the ability of a system to do something (work); we detect energy by the effect it has on the things around us, e.g. heating them up.

epicentre The point on the Earth's surface directly above the focus of an earthquake.

F

fibreglass A material containing extremely fine glass fibres which trap air – used in insulation.

fission Splitting apart, often used to describe the splitting of radioactive atoms such as uranium.

Fleming's left-hand rule The use of the first finger (field) and second finger (current) to predict the direction of movement of a conductor (thumb).

Fleming's right-hand rule The use of the first finger (field) and thumb (motion) to predict the direction of current in a conductor (second finger).

force A force is a push or pull which is able to change the velocity or shape of a body.

fossil fuel A fuel such as coal, oil and natural gas formed by the decay of dead animals and plants over millions of years.

free-fall A body falling through the atmosphere without an open parachute.

frequency The number of vibrations per second; frequency is measured in hertz (Hz).

friction A force that acts between two surfaces in contact with each other; it tends to prevent or slow down movement by the two surfaces.

fuse A special component in an electrical circuit containing a thin wire which is designed to heat up and melt if too much current passes through it, breaking the circuit.

G

galaxy A large, self-contained collection of stars and planets. Our galaxy is called the Milky Way.

gamma radiation Electromagnetic radiation emitted from the nucleus of a radioactive atom.

gamma camera A camera that takes pictures using gamma rays not light.

Geiger counter A device used to detect some types of radiation.

generator A device for converting movement (kinetic) energy into electricity.

gradient A slope – rate of change of quantity plotted on y-axis of graph compared to x-axis.

gravitational potential energy The energy a body has because of its position in a gravitational field, e.g. an object held above the ground has gravitational potential energy that can be released if the object is dropped.

gravity The force of attraction between two bodies caused by their mass; the force of gravity produced by a body depends on its mass – the larger the mass, the larger the force.

greenhouse gas Gases such as carbon dioxide and water vapour that increase the greenhouse effect.

H

hertz A unit of frequency equal to one cycle per second.

GLOSSARY

I

infrared Radiation beyond the red end of the visible spectrum. Infrared radiation is efficient at transferring heat.

insulation A substance that reduces the transfer of energy. Heat insulation in the loft of a house reduces the movement of warmth to the cooler outside. Sound insulation cuts down the reflection of sound in recording studios. Electrical insulation prevents the flow of electricity.

insulator A substance that will not let energy pass through it easily; you can have insulators for heat, electricity or sound.

interference Waves interfere with each other when two waves of similar frequencies occupy the same space. Interference occurs in light and sound and can produce changes in intensity of the waves.

ionisation The formation of ions (charged particles).

ionosphere A region of the Earth's atmosphere where ionisation caused by incoming solar radiation affects the transmission of radio waves; it extends from 70 kilometres to 400 kilometres above the surface.

J

joule A unit of energy.

K

kilowatt 1000 watts.

kinetic energy Energy due to movement.

L

laser A special kind of light beam that can carry a lot of energy and can be focused very accurately. Lasers are often used to judge the speed of moving objects or the distance to them.

latent heat The energy needed to change the state of a substance.

light A form of energy that allows us to see objects, light is given out by hot objects such as the Sun and the filament in an electric bulb.

light-emitting An electronic component that emits light when electricity passes through it.

logic circuits Circuits composed of a series of logic gates.

logic gates Electronic components that respond to signals by following preset logical rules.

longitudinal In longitudinal waves, the vibration is along the direction in which the wave travels.

M

magnet An object that is magnetic is attracted by a magnet.

magnetic field An area where a magnetic force can be felt.

mass Mass describes the amount of something; it is measured in kilograms.

melanin The group of naturally occurring dark pigments, especially the pigment found in skin.

meteorites A stony or metallic mass that has fallen to the Earth's surface from outer space.

methane A colourless, odourless gas that burns easily to give water and carbon dioxide.

molecule A group of atoms joined together by chemical bonds.

momentum The momentum of a body is equal to its mass multiplied by its velocity, it is usually measured in kilogram metres per second (kgm/s).

Morse code A code consisting of dots and dashes that represent each letter of the alphabet.

motor A device that converts electrical energy into movement energy.

mutation A random change in the genotype of an organism. Mutations are almost always disadvantageous.

N

National Grid The network of electricity cables that distribute electricity across the country.

neutral A material is neutral if it is either uncharged or has equal amounts of both positive and negative charge.

neutron A particle found in the nucleus of an atom, it has no electrical charge and a mass of 1 atomic mass unit.

newton The unit of force.

non-renewable Non-renewable fuels are not being made fast enough at the moment and so will run out at some point in the future.

nuclear To do with the nucleus.

nucleus The central part of an atom containing the protons and neutrons.

O

ohm The unit used to measure electrical resistance.

optical fibre A flexible optically transparent fibre, usually made of glass or plastic, through which light passes by successive internal reflections.

optics Anything to do with the behaviour of light rays.

orbit A path, usually circular, of a smaller object around a larger object, e.g. a planet orbits the Sun, electrons orbit the nucleus in an atom.

oscilloscope A device that displays a line on a screen showing regular changes (oscillations) in something. An oscilloscope is often used to look at sound waves collected by a microphone.

P

parabolic Shaped like a parabola, which looks a bit like an opened umbrella.

payback time The time taken to recover in cost savings the original outlay.

penetrating Passing into or through something.

photocell A device which converts light into electricity.

photon A unit or particle of electromagnetic energy. Photons travel at the speed of light but have no mass.

potential difference The difference in electrical potential energy between two points, measured in volts.

power The rate that a system transfers energy, power is usually measured in watts (W).

pressure The force acting on a surface divided by the area of the surface, measured in newtons per square metre (N/m^2).

projectile motion A particle direction, usually applied to the path of projectiles, e.g. bullets.

proton A particle found in the nucleus of an atom with a charge of plus one and a mass of one atomic mass unit.

R

radiation Energy that travels as a wave and requires no physical medium to carry it, e.g. light, radio waves or infrared radiation.

radio waves A form of electromagnetic radiation used to carry radio signals.

radioactive Material which gives out ionizing radiation.

radioactive waste Waste produced by radioactive materials used at power stations, research centres and some hospitals.

radiographer A technician who works in a hospital radiography department using X-rays to take radiographs.

radioisotope A radioactive isotope.

radiotherapy Using radiation to treat certain types of disease, e.g. cancer.

rarefaction A lowering of pressure caused by a sound wave passing a point.

reactor A device for maintaining a stable nuclear chain reaction.

reflection To bounce something back – usually light from a mirror or sound from a solid wall.

refraction A change in the direction of a light beam caused when the light crosses from one medium to another, e.g. when passing from air into a glass prism, the light beam seems to bend.

renewable Wind power, wave power and solar power are all examples of renewable sources of energy.

repel To push apart.

resistance The amount by which a conductor reduces an electric current.

resistor A conductor that reduces an electric current.

rheostat A variable resistor.

S

satellite A body orbiting around a larger body; communications satellites orbit the Earth to relay messages from mobile phones.

seismometer A device used to detect movements in the Earth's crust.

solar cell A device which converts light from the Sun into electricity.

Solar System The collection of eight planets and other objects orbiting around the Sun.

specific heat capacity The amount of energy needed to raise the temperature of 1 kg of a substance by 1 degree Celsius.

specific latent heat The amount of energy needed to change the state of 1 kg of a substance without changing its temperature, e.g. the energy needed to change 1 kg f ice at 0 °C to water at the same temperature.

speed How fast an object is moving.

static electricity Electric charge on the surface of a non-conductor, i.e. it is charge that cannot flow easily.

sterilise Killing all the organisms in an area, usually used to mean killing microorganisms.

stopping distance The total distance required to bring a moving vehicle to a halt.

stratosphere A layer in the atmosphere starting at 15 km above sea level and extending to 50 km above sea level. The ozone layer is found in the stratosphere.

T

terminal speed The speed at which the force of gravity and the force due to air resistance are equal and the object is falling as fast as possible.

terminal velocity The velocity at which the force of gravity and the force due to air resistance are equal and the object is falling as fast as possible.

thermal energy Energy that can raise the temperature of an object, sometimes called 'heat'.

tracer A radioactive, or radiation-emitting, substance used in a nuclear medicine scan or other research where movement of a particular chemical is to be followed.

trajectory The path a projectile follows.

transformers Devices which can change the size of an AC voltage and current.

transmitter A device which gives out some form of energy or signal, usually used to mean a radio transmitter which broadcasts radio signals.

transverse In transverse waves, the vibration is at right angles to the direction in which the wave travels.

turbine A device that converts movement in a fluid into circular movement, usually to drive a generator. Turbines are essential parts of a windmill and a hydroelectric power plant.

GLOSSARY

U

ultrasound Sounds which have too high a frequency for humans to hear (above 20 kHz).
ultraviolet Radiation just beyond the blue end of the spectrum of visible light. UV light is important in tanning and some sorts of skin cancer.
Universe Everything, everywhere.
uranium A radioactive metal used in nuclear power stations and bombs.

V

variable This is something that can change if you were doing an investigation.
variable resistor A device whose resistance can change, often used in volume controls and light dimmer switches.
vector Physical quantity that has both magnitude and direction.
velocity Velocity is the speed an object is moving in a particular direction; a change in direction or speed will change the velocity; velocity is usually measured in metres per second (m/s).
volt The unit of electrical potential.
voltage The potential difference across a component or circuit.
voltmeter A meter that measures the voltage between two points.

W

watt A unit of power, 1 watt equals 1 joule of energy being transferred per second.
wavelength The distance between two successive identical points of displacement on a wave.
waves A ripple or undulation; all electromagnetic radiation, including radio signals, light rays, X-rays, and cosmic rays, as well as sound, behave like rippling waves in the ocean.
weight The force of gravity acting on a body on Earth; weight is a force and is measured in newtons.
work Work is done when a force moves; the greater the force or the larger the distance, the more work is done.

X

X-ray Electromagnetic radiation used by doctors to look inside a patient's body or to destroy some types of cancer cells.

Answers

Here are the answers to the topic questions from the revision guide pages, and the answers to the exam-practice questions in the workbook.

This section is perforated, so that you can remove the answers to help test yourself or a friend.

P1 Energy for the home
Page 4
1. Temperature is a measure of hotness, heat is a form of energy.
2. Energy is transferred from your hand to the ice.
3. It changes into a liquid.
4. The specific heat capacity of the syrup is greater than the specific heat capacity of the sponge.

Page 5
1. Fibreglass contains trapped air.
2. It reflects radiation back into the room.
3. 60 years.
4. £97.50.

Page 6
1. Air trapped in the fibres acts as an insulator.
2. Cooking and producing electricity.
3. A vacuum does not conduct heat.
4. Less energy is transferred through the ceiling into the roof.

Page 7
1. An electric fire.
2. For safety - microwaves could warm the water in human body.
3. Do not use mobile phones unless they have to or send text messages.
4. They are not in line of sight; they are in valleys, behind hills or tall buildings.

Page 8
1. Infrared.
2.
3. Any two from: visible light, infrared, laser.
4.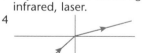

Page 9
1. It can be used anywhere.
2. Less refraction of other waves that could interfere.

Page 10
1. 300 000 km/s.
2. 300 m/s.
3. It is quicker.
4. It uses a series of dots and dashes to represent letters of the alphabet.

Page 11
1. An instrument for detecting earthquakes.
2. The focus is the source of an earthquake. The epicentre is the point on the Earth's surface above the focus.
3. Burning fuels, cutting down trees.
4. 150 minutes / 2.5 hours.

P2 Living for the future
Page 13
1. Photo cells use light and solar cells use light from the Sun.
2. No mains electrical supply is needed in remote areas.
3. So that it absorbs more radiation.
4. Sun over the equator – the equator is north of Australia but south of England.

Page 14
1. a DC b AC.
2. Coal, oil, natural gas.
3. Steam is pressurised.
4. There is less energy loss and lower distribution costs.

Page 15
1. An energy source which is being used up faster than it can be produced.
2. No carbon dioxide or smoke is produced.
3. Radiation causes ionisation, which changes the structure of atoms. DNA in the cell can change, so the cell behaves differently and divides in an uncontrolled way. This causes cancer.
4. 5p

Page 16
1. Radiation that is always present.
2. 75%.
3. It increases the distance from the radioactive source and there is no skin contact.
4. Alpha radiation would not penetrate. Gamma radiation would give little change in the count rate when thickness changes slightly.

Page 17
1.
2. They fluoresce.
3. A body in orbit around another body.
4. From a collision between two planets.

Page 18
1. Moon, planets, artificial satellites.
2. Mercury, Venus, Earth, Mars, Jupiter, Saturn, Uranus, Neptune, (Pluto).
3. 20 000 years.
4. To keep the body at a suitable temperature, maintain suitable pressure, provide oxygen for breathing.

Page 19
1. Fires, tsunamis, dust clouds, falling temperature, extinction of life forms.
2. Large rocks or small planets orbiting the Sun.
3. Dust and rocks.
4. In case their orbit passes close to the orbit of Earth.

Page 20
1. Hydrogen and helium.
2. The galaxies furthest away.
3. Light cannot escape from it.
4. The remnants from a large star that has become a supernova can merge to form a new star.

P3 Forces for transport
Page 22
1. m/s or km/h.
2. 30 m/s.
3. a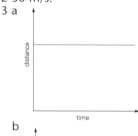
   b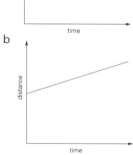
4. a It is constant
   b It is decreasing.

Page 23
1.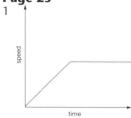
2. The distance travelled.
3. It is decelerating.
4. 2.5 m/s^2.

Page 24
1. Car B.
2. 6000 N.
3. Braking distance.
4. 14 m.

Page 25
1. Ali.
2. 1600 J.
3. Meera weighs more than Jan.
4. 400 W.

Page 26
1. a Its food
   b coal or gas
   c falling water.
2. A larger mass means the car must gain more kinetic energy, so it uses more fuel.
3. The diesel-driven car.
4. One from: short range, slow speed, batteries take time to recharge, need a recharging facility, batteries take up a lot of space/are heavy, etc.

Page 27
1. Any two from: thermal energy, elastic energy, sound energy.
2. They open and close quickly at the push of a button so the driver is not distracted for too long and can concentrate on driving.
3. The seat belt and air bag.
4. This is less tiring on long motorway journeys. It helps a car keep to a steady speed so the driver does not have to use the pedals.

Page 28
1. A feather has a bigger surface area, so there is a bigger air resistance force; smaller weight.
2. Weight, air resistance / drag.
   a Weight constant, drag = 0
   b Weight = drag c drag greater than weight.
3. It increases the car's top speed.
4. There are more air molecules displaced each second.

Page 29
1. Jo's.
2. Gravitational PE > KE > gravitational PE, etc.
3. So less energy is needed to raise it / it needs to gain less gravitational PE.

4 It has very little KE at the top (mainly gravitational PE) but a lot of KE at the bottom.

P4 Radiation for life
Page 31
1 b Paper c polythene.
2 a They repel b they attract.
3 Clothes are insulators and rub against each other, becoming charged by friction.
4 They could cause a spark which may lead to an explosion as fuel is highly inflammable.

Page 32
1 To ensure good electrical contact.
2 Negative.
3 To charge the soot particles when they come near the wires.

Page 33
1 It decreases.
2 8 Ω.
3 The earth wire.
4 The wire in the fuse melts if the current becomes too large, breaking the circuit and preventing overheating.

Page 34
1 256
2 Body fat.
3 It avoids major surgery.
4 10 cm.

Page 35
1 Both are very penetrating and can pass into the body.
2 Using radiation to treat diseases such as cancer.
3 It kills bacteria.
4 It emits only gamma radiation or iodine is taken up by the thyroid gland.

Page 36
1 600 Bq.
2 Radioactive decay is a random process.
3 a A helium nucleus b an electron.
4 It is radiation, not a particle.

Page 37
1 To locate a leak or blockage in a buried pipeline.
2 So that a decrease in ionisation current is due to the presence of smoke and not to the decay of the source reducing the number of ions present.
3 Less uranium has decayed to lead (i.e. more uranium and less lead than an old rock).
4 Iron was never living so does not contain carbon-14.

Page 38
1 Uranium.
2 It is uranium containing a greater proportion of the uranium-235 isotope than occurs naturally.
3 In a nuclear reactor.
4 It emits harmful ionising radiation for a long time.

P5 Space for reflection
Page 40
1 Ganymede.
2 Earth takes 24 hours to rotate on its axis.
3 Spying – military use; research – space station / exploration; mapping – monitoring Earth surface / Ordnance Survey; GPS – location / satnav.
4 Polar orbit is lower than geostationary orbit.

Page 41
1 240 km/h.
2 Scalar – mass; temperature; volume. Vector – force; acceleration.
3 10 m/s.
4 45 m.

Page 42
1 Dart; javelin; discus; shot put; hammer; caber; any ball, e.g. foot, golf.
2 Zero.
3 0.459 N.

Page 43
1 Equal and opposite force on first child.
2 4.5 kgm/s.
3 Boot + ball; ball + head; player + player; ball and goalpost.
4 Increases time to stop, so less force.

Page 44
1 Lower frequency radio waves will not pass through atmosphere – geostationary satellite is above atmosphere.
2 Lower frequencies reflected back, higher frequencies scattered.
3 Reflected.
4 To make them in line of sight.

Page 45
1 Waves spread out and interfere.
2 Trough and crest overlapping.
3 Sharp shadows same shape as object.
4

Page 46
1 Red.
2 Increases.
3 Water is the more dense medium.

4

Page 47
1 Eye; spectacles; headlamp; lighthouse.
2 Light is a parallel beam and parallel to the principal axis.
3 More pixels means better definition.
4 Can focus on close and distant objects to give sharp image.

P6 Electricity for gadgets
Page 49
1 Dimmer switch, Scalextric controller.
2 Decreases/dimmer.
3 Gets so hot it melts.
4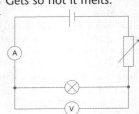

Page 50
1 Control on hi-fi / cooker / radio / etc.
2 0 V; 6 V; 12 V.
3 Resistance.
4 Resistance is high when it is dark.

Page 51
1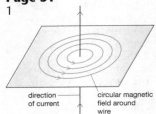

2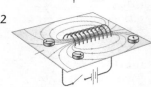

3 Turntable of microwave / beaters of blender / any correct application.
4 Deliver current to commutator.

Page 52
1 A current is produced in the coil.
2 To the left.
3 Current increases; frequency is higher.

Page 53
1 Step-down.
2 11.5 V.
3 Step-up.
4 9 times.

Page 54
1 John's diode is connected the opposite way round to Jane's.
2
3 Store charge.
4 a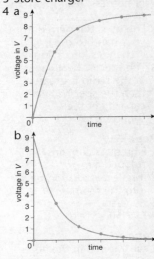
b

Page 55
1
input X	input Y
0	1
1	0

2
input X	input Y	output Z
0	0	0
0	1	0
1	0	0
1	1	1

input X	input Y	output Z
0	0	0
0	1	1
1	0	1
1	1	1

3 Resistance is high when cold and low when hot.

Page 56
1 An electronic system is a combination of logic gates (and other devices).

2
A	B	C	D	E
0	0	0	0	0
0	0	1	0	0
0	1	0	1	0
0	1	1	1	1
1	0	0	1	0
1	0	1	1	1
1	1	0	1	0
1	1	1	1	1

3 The current from a logic gate is too small to operate a bulb.
4 A magnetic field is produced around the coil.

Workbook answers

Remember: Check which grade you are working at.

P1 Energy for the home

Page 62 Heating houses
1. **a** Degrees Celsius; energy; joules
   b Energy flows from warm to cooler body; temperature of warmer body falls
2. **a** 8 minutes
   b Iron is a different material
   c Specific heat capacity
3. **a** It changes state from solid to liquid
   b It changes state from liquid to gas
   c Melting ice; boiling water *(Any 1)*
   d Specific latent heat

Page 63 Keeping homes warm
1. **a** Air is a good insulator/a poor conductor
   b Cavity wall insulation; drawing curtains; sealing gaps; shiny foil behind radiators; carpets/underlay *(Any 3)*
   c 120 ÷ 40 = 3 years
   d Shorter payback time
   e Only 32% of energy input is useful; as energy output
   f Energy is lost up the chimney

Page 64 How insulation works
1. Trapped; insulator
2. **F** in bottom left-hand corner of room.
3. **a** Particles in solid close together; gap between glass filled with gas/vacuum; particles in gas far apart/no particles in vacuum; more difficult to transfer energy than in solid *(Any 3)*
   b i Air in foam is good insulator; reduces energy transfer by conduction; air is trapped; unable to move; reduces energy transfer by convection *(Any 4)*
   ii Energy from room reflected back into room in winter; energy from Sun reflected back outside in summer

Page 65 Cooking with waves
1. **a** Infrared; absorb; spectrum; water
   b Microwave radiation is more penetrating than infrared; microwave ovens cook by conduction and convection *(Any 1)*
2. **a** Some evidence of heat energy being transferred to the body; young people more likely to be affected
   b Microwaves need line of sight; no obstructions in space
   c Amplified; re-transmitted back to Earth

Page 66 Infrared signals
1. **a** True; false; true   **b** Digital
   c
2. **a**
   b i (i) (ii) (iii)
   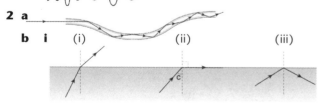
   ii *(See diagram)*

Page 67 Wireless signals
1. **a** Can be used anywhere/portable   **b** Reflected
   c The radio station is broadcasting on the same frequency; the radio waves travel further because of weather conditions
   d Atmosphere; frequency

Page 68 Light
1. **a** Radio; microwave; infrared; ultraviolet; X-rays; gamma ray *(Any 4)*
   b *(See diagram)*
2. **a** *(See diagram)*   **b** *(See diagram)*
   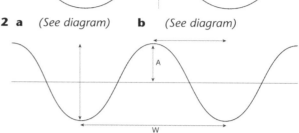
   c Number of complete waves passing a point each second
3. **a** Semaphore; smoke signals; runner *(Any 1)*
   b Need to represent letters as electric signal

Page 69 Stable Earth
1. **a** Seismometer
   b Fault; shock
   c

description	P wave	S wave
pressure wave	✓	
transverse wave		✓
longitudinal wave	✓	
travels through solid	✓	✓
travels through liquid	✓	

2. **a** Climate change
   b Carbon dioxide
   c Reflects radiation back down to Earth
   d i Sunburn; skin cancer *(Any 1)*
   ii Can stay in sun 30 times longer without burning

P2 Living for the future

Page 71 Collecting energy from the Sun
1. Light for photosynthesis; heat for warmth
2. **a** Light
   b Robust/not much maintenance; no fuel/no power cables; no pollution/no contribution to climate change; renewable energy source *(Any 4)*
   c Less electricity produced
3. **a i** Black absorbs more energy   **ii** Convection
   b Walls and floors radiate energy back into room

Page 72 Generating electricity
1. **a** Ammeter needle moves to left
   b Stronger magnet; more turns on coil; turn coil faster *(Any 2)*
2. Magnetic field turns inside coil of wire
3. **a** Turbine turns generator
   b Wires become warm and lose energy to surroundings
4. **a** Change size of AC voltage
   b More energy would be lost as heat from transmission wires

Page 73 Fuels for power
1. **a** Straw; wood
   b i Substance that burns; releases energy as heat
   ii Atoms of uranium split

WORKBOOK ANSWERS

2 a Radioactive nuclear waste produced
b Nuclear; DNA; mutate; cancer
3 a Power = voltage × current
= 12 × 2
= 24 W
b Cost = power × time × cost per kWh
= 2.5 × 0.5 × 12
= 15p

Page 74 Nuclear radiations
1 a Geiger counter
b Radon gas; rocks and soil; cosmic rays; food and drink (Any 2)
2 False; true; true; false
3 Safe distance; do not handle directly; shield; use for minimum time; do not point at his body (Any 3)
4 a Kills living cells/bacteria
b i Tracer
ii Penetrating
c Wall is thinner

Page 75 Our magnetic field
1 a (N at bottom of magnet = 1 mark)
b Magnetic field is similar to a bar magnet

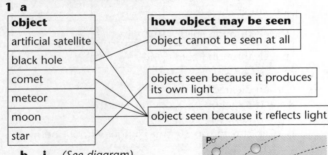

2 Navigation/GPS; spying; communication; mapping; weaponry (Any 2)
3 a Communication signals disrupted
b Charged particles
4 The presence of the Moon

Page 76 Exploring our Solar System
1 a

object	how object may be seen
artificial satellite	object cannot be seen at all
black hole	
comet	object seen because it produces its own light
meteor	
moon	object seen because it reflects light
star	

b i (See diagram)
ii (See diagram)
2 a Sending radio signals
b Always gravitational forces acting however small
c Avoid being blinded by Sun

Page 77 Threats to Earth
1 a It got colder; there were lots of fires; tsunamis flooded large areas
b Mars; Jupiter
c Rocks left over when the Solar System formed
2 a

b When it is near the Sun
c Solar winds blow dust
3 a Near–Earth–objects/comets or asteroids whose orbits pass close to the orbit of Earth
b Increased accuracy of predicting if a collision may occur

Page 78 The Big Bang
1 a A single point
b Helium; hydrogen; protons
c Those furthest away

2 a Cloud of gas and dust
b i Red giant; core contracts; outer changes colour from yellow to red expands; planetary nebula thrown out; core becomes white dwarf; cools to become black dwarf (Any 4)
ii Red supergiant; core contracts; outer expands; core becomes neutron star; explosion/supernova; neutron star becomes black hole; supernova remnants become new stars (Any 4)

P3 Forces for transport
Page 80 Speed
1 a i Eve
ii Eve covered the same distance as Melissa but in a shorter time
b i Distance travelled; time taken
ii Distance: meter rule/tape; time: stop clock; light gates with accurate timer; data logger
c Distance = 8 × 1.5 = 12 m
Time = 0.5 s
Speed = distance travelled in 1 s = 12 × 2 = 24 m/s
d i Average speed = $\frac{distance}{time}$
= $\frac{390}{3}$
= 130 km/h
ii Car cannot maintain the same speed throughout
iii Yes; if average speed is 130 km/h the car must have gone faster (and slower) than this at times
2 a A to B; E to F **b** 80 s **c** 80 m **d** D to E

Page 81 Changing speed
1 a i Increases steadily from rest; reaching a speed of 10 m/s after 30 s; then travels at a constant speed of 10 m/s
ii

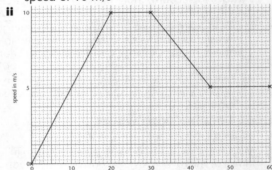

iii Area under graph
b i

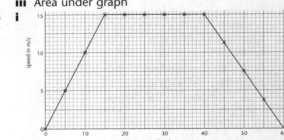

ii Yes; 50 km/h = 50 000 m/h
= $\frac{50\,000}{3600}$ m/s
= 13.9 m/s
2 Acceleration = $\frac{change\ in\ speed}{time}$
= $\frac{(40-10)}{6}$
= 5 m/s^2

Page 82 Forces and motion
1 Accelerate
b i Increase
ii Greater acceleration
c Smaller acceleration

2 a $a = \dfrac{(v-u)}{t}$

$= \dfrac{(40-0)}{20}$

$= 2$ m/s^2

b $F = ma$
$= 500 \times 2$
$= 1000$ N

c Resistive forces ignored

3 a Distance travelled by car between applying the brakes and the car stopping

b Distance travelled by car between seeing the need to brake and applying the brakes (time for brain to react)

c Stopping distance = thinking distance + braking distance

d Increase; thinking/reaction time unchanged; but she will travel a greater distance in that time at a higher speed

e Tired; under influence of alcohol/other drugs; distracted/lacking concentration *(Any 2)*

Page 83 Work and power

1 a More **b** Less **c** None

d The food that she eats

e WD = force × distance = 80 × 2 = 160 J

2 a Same **b** Chris is more powerful than Abi

c 60 × 10 = 600 N

d WD = force × distance = 600 × 3 = 1800 J

e Chris' power = $\dfrac{\text{WD}}{\text{time taken}}$

$= \dfrac{1800}{8}$

$= 225$ W

f Abi's power = $\dfrac{1800}{12}$

$= 150$ W

3 a A **b** C

c Fuel pollutes the environment; car exhaust gases are harmful; carbon dioxide is a major source of greenhouse gases; carbon dioxide contributes to climate change *(Any 3)*

Page 84 Energy on the move

1 a Energy of an object that is moving **b** B; C

2 a i 1.6 litre diesel

ii Diesel

iii Higher mpg than both petrol cars

b i Fewer road junctions; speed changes; gear changes

ii $\dfrac{96}{24} = 4$

iii Less

iv Land Rover has a bigger engine capacity

3 a Recharging requires electricity from power stations which cause pollution

b i Advantage: energy from Sun so causes no pollution; do not have batteries that need recharging; do not use electricity from power stations *(Any 1)*

ii Sun does not always shine; does not have a constant energy source *(Any 1)*

Page 85 Crumple zones

1 a i Crumple zones **ii** Air bag **iii** Seat belt

b

safety feature	how it works
seatbelt	designed to stretch a little so that some of a person's KE is converted to elastic energy
crumple zones	absorb some of the car's KE by changing shape (crumpling) on impact
air bag	absorbs some of a person's KE by squashing up around them

c i Directly improve the safety of a car

ii Indirectly improve the safety of a car

d i ABS brakes; traction control; safety cage *(Any 2)*

ii Electric windows; cruise control; paddle shift controls; adjustable seating *(Any 2)*

e i To be effective they need to be kept in good working order

ii Car crash may have damaged some of the safety features

iii In case the belt fabric has been overstretched

f i Anti-lock braking system

ii Driver gets maximum braking force without skidding; can still steer car

Page 86 Falling safely

1 a Gravity pulls them towards the Earth

b Increases

c Golf ball heavier than ping pong ball; so air resistance has a bigger effect on ping pong ball

d Golf ball

e i 10 m/s^2

ii (See diagram)

iii Weight greater than air resistance

iv The faster she falls the more air molecules she displaces each second; so the greater the air resistance force; net/resultant force is less so acceleration is less

v Terminal speed

vi Balanced; equal in size but opposite in direction

2 a To reduce drag; air resistance force acting on them; slowing them down *(Any 2)*

b Crouch over handlebars; wear tight-fitting clothes; wear shaped helmet with no sharp edges and sleek shape; lubricate bicycle; have bicycle with no sharp edges and sleek shape *(Any 2)*

Page 87 The energy of games and theme rides

1 GPE; GPE; KE; GPE

2 a A: GPE; B: GPE + KE, C: EPE (elastic potential energy)

b Some energy is converted into thermal energy and sound

3 a B **b** Increases **c** C **d** GPE to KE

e Energy is transferred to other forms: sound/thermal energy; due to friction

f Make B higher so it gains greater GPE; GPE lost = KE gained + energy transferred due to friction; its maximum KE will be greater; more KE means higher speed *(Any 3)*

P4 Radiation for life

Page 89 Sparks!

1 a Negative

b i Acetate/perspex

ii It will pick up the pieces of paper

c Copper is an electrical conductor so charges will not stay on it

2 a So that charge cannot pass through her

b So that she does not get an electric shock

c i Sally becomes charged

ii All her hairs gain the same charge; like charges repel so the hairs move away from each other

3 a The car becomes charged due to friction with the air on the journey; you are not charged so charge flows through you when you touch the car door

b Lightning may strike the tree as it is the tallest object around

c Cling film becomes charged due to friction as it is unrolled (as electrons are transferred from one part of the film to another, areas acquire opposite charges, so attract)

WORKBOOK ANSWERS

d Nylon is an electrical insulator; Priya becomes charged by friction as she walks

e Bare wire is highly charged; when Tom touches it charge flows through him to Earth (giving a serious electric shock)

Page 90 Uses of electrostatics

1 a To re-start a heart
b Makes the heart contract
c Place paddles firmly on chest; no clothes/hairs
d It only passes for a very short time

2 a Even coverage; less paint wasted; paint covers awkward places *(Any 1)*
b Paint droplets all have the same charge; like charges repel
c To attract the paint droplets to the frames; as opposite charges attract

3 a To remove harmful particles that pollute the atmosphere
b Fossil fuel power stations
c Positive

Page 91 Safe electricals

1 a i Circuit incomplete
 ii (Complete circuit = 1 mark)
b i Varying brightness
 ii Ammeter in series; voltmeter in parallel with lamp
 iii 24 ohms

2 a Live; neutral; earthed; power station
b Battery is DC, mains is AC; battery is lower voltage than mains

3 a (Live: on right (to fuse); neutral: on left; earth: at top)
b Live
c i (See diagram)
 ii Breaks circuit if a fault occurs
d i Earth
 ii It is connected to the metal case of an appliance to prevent it becoming charged if touched by a live wire; it provides a low resistance path to the ground
e 13 A

Page 92 Ultrasound

1 a Vibrations are in same direction as the wave
b The number of vibrations in a second
c i Vibrations set up pressure wave in air – compressions (higher pressure) and rarefactions (lower pressure); make eardrum vibrate
 ii Frequency increases
 iii Sound of a higher frequency than humans can hear

2 a To check the condition of the foetus
b 1 000 000 Hz (or 1 MHz)
c Measure speed of blood flow; clean teeth/old buildings/jewellery; break down stones in the body *(Any 2)*
d Pulse; tissues; reflected; echoes; image; gel; probe; skin; ultrasound/pulse; reflected; skin
e i Very rapid ultrasound vibrations break the stones down into small pieces that are excreted from the body in the normal way
 ii It needs to be powerful enough/carry enough energy; to break up the stones

Page 93 Treatment

1 a Diagnosis: finding out what is wrong with a patient; therapy: treatment
b Similarity: both electromagnetic radiation (of very short wavelength); difference: gamma rays emitted from the nucleus of an atom, X-rays are not (produced in an X-ray machine)
c Both very penetrating/can pass into the body to treat internal organs
d Alpha particles cannot penetrate skin; beta particles would be stopped by a small thickness of tissue and by bone

2 a It damages and destroys cancerous cells
b Destroying cancerous cells by exposing the affected area of the body to large doses of radiation
c To make sure all the cancerous cells are removed (by surgery) or destroyed (by radiotherapy)
d To sterilise equipment

3 a A tiny amount of a radioisotope introduced into the body
b To investigate a problem without surgery
c Occasionally beta but usually gamma radiation
d Thyroid
e X-rays are produced in an X-ray tube; gamma rays can be emitted inside the body and their progress monitored

Page 94 What is radioactivity?

1 a $\dfrac{\text{Average number of nuclei that decay}}{\text{time taken}}$
b i Geiger-Muller tube
 ii Ratemeter
c The decay of one nucleus
d i Activity = $\dfrac{\text{number of nuclei that decay}}{\text{time taken in s}}$
 = $\dfrac{750}{30}$
 = 25 Bq
 ii Decrease
 iii Radioactive decay is a random process; all experimental results should be repeated if possible

2 a

type of radiation	charge (+, – or 0)	what it is	particle or wave
alpha	+	helium nucleus	particle
beta	–	electron	particle
gamma	0	electromagnetic radiation	wave

b Gamma; alpha; alpha; gamma; beta

Page 95 Uses of radioisotopes

1 a Ionising radiation; that is always present in the environment
b Radioactive substances present in rocks (especially granite) and soil; cosmic rays from space

2 Detect leaks in underground pipes; monitor the uptake of fertilisers in plants; check for a blockage in a patient's blood vessel; track the dispersal of waste material; track the route of underground pipes *(Any 3)*

3 a It is highly ionising/short range in air
b Alpha particles ionise atoms in the air; + and – ions move towards – and + plates respectively; this creates a tiny current which is detected; if present smoke particles attach themselves to the ions neutralising them; current falls setting off alarm

4 a Very little change in count rate over 200 year period
b Seeds; animal bone
c Rocks never contained living matter

Page 96 Fission
1 **a**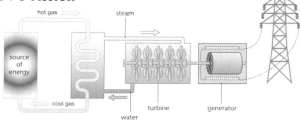
   b Source of energy; water; steam; steam; turbine; generator
2 **a** **i** Uranium
     ii The splitting of a large nucleus such as uranium with the release of energy
   b **i** Chain reaction
     ii Chain reaction controlled in nuclear power station, but is out of control in a nuclear bomb
3 **a** **i** Put the materials in a nuclear reactor
     ii To produce artificial radioisotopes; in hospitals to diagnose/treat patients; in industry as tracers to detect leaks (*Any 1*)
   b **i** An uncharged particle found in the nucleus of an atom
     ii Uncharged; so can penetrate deep inside nucleus easily; producing unstable isotope
4 **a** Embedded in glass discs and buried in the sea/incinerated under strict controls (very low waste only) (*Any 1*)
   b Reprocessed

P5 Space for reflection
Page 98 Satellites, gravity and circular motion
1 **a** Moon
   b Planets
   c Amplified; re-transmitted
   d 24 hours
   e Communications
   f View whole of Earth's surface
2 **a** Gravity
   b Force acting towards centre of circle; keeping body moving in a circle

Page 99 Vectors and equations of motion
1 **a** 20 mph
   b Scalar has size only; vector has size and direction
   c 1050 N
   d 0 N
2 **a** Not always travelling at maximum speed
   b Speed = distance ÷ time; = 186 ÷ 3; 62 mph
   c $v = u + at$; $v = 9 + (3.6 \times 5)$; 27 m/s

Page 100 Projectile motion
1 **a** Projectile; trajectory; parabola
   b **i** Exactly 2.4 s
     ii 4 m/s
   c Fire arrow at an angle above the horizontal but not above 45°
2 Ball falls to Earth; but Earth curves away beneath ball

Page 101 Momentum
1 **a** Action; reaction; 60 N
   b Force of attraction towards Earth
2 **a** When one body hits another
   b **i** Momentum = mass × velocity; = 75 × 9; 675 kgm/s
     ii Force = change in momentum ÷ time; = 675 ÷ 0.5 or 1350 N (allow error carried forward from **i**)
     iii Force would be greater (*1 mark*); force would be ten times greater or 13 500 N (*2 marks*)
     iv Force would be greater (*1 mark*); force would be three times greater (*2 marks*)

Page 102 Satellite communication
1 **a** 1000 km; 1 000 000 km
   b Absorbed; reflected (*Any order*)
   c Detect radio waves
   d Parabolic receiver
   e Amplified; re-transmitted back to Earth
2 **a** Microwaves have a shorter wavelength
   b Spreading out of a wave as it passes through a gap or around an obstacle

Page 103 Nature of waves
1 **a** Louder, brighter, quieter, darker
   b Sound becomes louder and quieter; and louder again
2 **a** Laser
   b Straight
   c Alternate bands/stripes/bars/fringes of red light; and darkness
   d Sound is not a transverse wave; only transverse waves can be polarised

Page 104 Refraction of waves
1 **a** Refraction
   b Dispersion
   c 200 000 km/s
   d Sunlight is made up of different colours; each has different refractive index so refract by different amounts
2 **a** Ray is totally reflected back inside the plastic
   b Endoscope; to see inside body without surgery; communications; transmission of laser light signals

Page 105 Optics
1 **a** Convex; refracted; focus
   b Real images can be projected onto a screen
   c Projector image is larger than object/camera image is smaller than object
2 **a** Magnifying glass

part	job
curved mirror	reflects light back through the film
condenser lenses	concentrate light onto the film
projection lens	focuses image onto the screen

P6 Electricity for gadgets
Page 107 Resisting
1 **a** Amp; resistance; volt
   b Resistance increases
   c $R = V \div I$ or $230 \div 0.5$; 460 Ω
2 **a** Ammeter in series; voltmeter in parallel

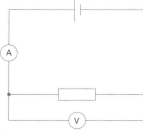

   b Hot; increase
   c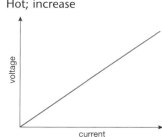

Page 108 Sharing

1 a Voltage; resistance; divider
  b Change R_1 / R_2 into a variable resistor
2 a Heat
  b Resistance of a thermistor decreases with temperature
  c Resistance of LDR changes with light intensity; used to automatically switch on street light

Page 109 Motoring

1 a Current produces magnetic field; fields interact; repulsion
  b Moves in opposite direction
  c Moves in opposite direction
  d Field pattern; direction of field relative to current

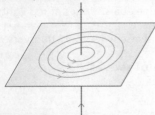

2 a Drill; screwdriver; lawnmower; food mixer; food blender; microwave turntable; fan; record deck; electric can opener (Any 1)
  b Increase current; increase number of turns on coil; increase magnetic field

Page 110 Generating

1 a Alternating
  b The reading remains at zero
  c A current is produced
  d Current in opposite direction
2 a Lamp goes out
  b Direct current supplied; to coils around rotor coil
  c Maintain constant voltage; constant frequency

Page 111 Transforming

1 a Increase the voltage
  b Shaver socket in bathroom
  c i Step-down
    ii $V_p = V_s n_p \div n_s$; 12 × 10 000 ÷ 500; 240 V
2 a National Grid
  b 400 000 V
  c Hot; energy; four

Page 112 Charging

1 a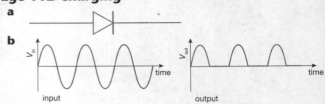
  b
  c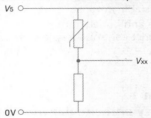
2 a Store charge
  b Smoothes output to make it more constant

Page 113 It's logical

1 a 5; high; 1; low; 0
  b There is no output / output is low / 0
  c OR; AND
2 Potential divider circuit; thermistor in correct place

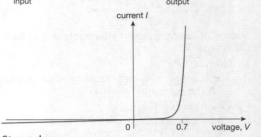

Page 114 Even more logical

1 a System
  b

A	B	C	output
0	0	1	1
0	1	1	1
1	0	0	0
1	1	0	1

2 a Current from logic gate too low to light bulb
  b Relay
  c Indicator to show something is working